RICHARD GIRLING

The Man Who Ate the Zoo

Frank Buckland, forgotten hero
of natural history

VINTAGE

For Oscar, Leo and Bonnie

1 3 5 7 9 10 8 6 4 2

Vintage
20 Vauxhall Bridge Road,
London SW1V 2SA

Vintage is part of the Penguin Random House group of companies
whose addresses can be found at global.penguinrandomhouse.com

Penguin
Random House
UK

First published in Vintage in 2017
First published in hardback by Chatto & Windus in 2016

penguin.co.uk/vintage

A CIP catalogue record for this book is
available from the British Library

ISBN 9781784701611

Printed and bound by Clays Ltd, St Ives Plc

Penguin Random House is committed to a sustainable future
for our business, our readers and our planet. This book is made
from Forest Stewardship Council® certified paper.

RICHARD GIRLING

Richard Girling is an award-winning environmental journalist. For his work in the *Sunday Times* he has been named Specialist Writer of the Year in the UK Press Awards. He was Journalist of the Year at the Press Gazette Environmental Press Awards in 2008 and 2009. His most recent book, *The Hunt for the Golden Mole*, was highly praised: 'This is a book that bursts into life from the first page . . . Rousing, fascinating . . . Utterly engaging.' *Sunday Times*

ALSO BY RICHARD GIRLING

Fiction
Ielfstan's Place
Sprigg's War

Non-fiction
*The View from the Top: A Panoramic Guide to Reading
Britain's Most Beautiful Vistas*
Rubbish!: Dirt on our Hands and Crisis Ahead
Sea Change: Britain's Coastal Catastrophe
Greed: Why We Can't Help Ourselves
*The Hunt for the Golden Mole: All Creatures Great
and Small, and Why They Matter*

CONTENTS

Introduction

Here is the beginning.
The first recorded words of Francis Trevelyan Buckland,
aged not quite four: 'The vertebrae of an ichthyosaurus.'
And the end.
The last recorded words of Frank Buckland, Her Majesty's
Inspector of Salmon Fisheries, aged fifty-four years and
two days, December 1880: 'I am going on a long journey
where I think I shall see a great many curious animals. This
journey I must go alone.'

In death as in life, Frank Buckland travelled expectantly.
For him it was the only way. To anticipate *curious animals*,
and to anticipate them in *heaven*, was an automatic reflex for
a man who had devoted his life to them. Every waking hour,
from dawn until deep into the candled night, he had pored
over fur, feather and fin. His absorption was both intellectual and physical. Intellectual because he wanted to know why
things happened. Physical because no enquiry was complete
until he had stripped an animal from its bones, weighed its
brain, measured its lungs, blown smoke through its cavities
and – the root of his notoriety – found what it tasted like.
Dog and cat, mouse and rat, panther, rhino, giraffe . . . dainty
morsels all. The only things that disgusted him were earwigs
and cucumber.

The world, then and since, has never been quite sure what to
make of Frank Buckland. Zoologist, anatomist, surgeon, pisci-
culturist, bestselling author, sell-out lecturer, lover of freak

shows, friend of the famous, royal consultant, joker, zoophagist, official purifier of rivers, saviour of the seas, curator, public benefactor, hypocrite, force of nature, buffoon. Insofar as he is now remembered at all, it is as a five-star exhibit in the grand pantheon of Victorian eccentricity. The man who ate everything. The man who thought he was a salmon. The man who cut up his father and made a public exhibition of his bones. What larks! Frank was born and died a brilliant child, his life a glorious passage of uninterrupted boyhood. His zeal was exultant, unquenchable, unapologetic, unrestrained by taboos. Nearly everybody loved him.

He has had two previous biographers, and I have rather surprised myself by becoming the third. I began with the intention of writing about his friend, the taxidermist turned greatest ever superintendent of London Zoo, Abraham Dee Bartlett. Frank turned up in Bartlett's correspondence, and curiosity then sidetracked me onto his books. I wrote a little about him in my recent book *The Hunt for the Golden Mole*:

> When an old lion died at the zoo, he was present at the dissection to peel the skin from the foot and fiddle with the tendons (they worked 'with the ease of a greased rope in a well-worn pulley'). When lions broke out of their cage at Astley's Royal Amphitheatre in Westminster Bridge Road, Lambeth – birthplace of the circus ring – he was on hand to examine the corpse of the unlucky stable-hand who got in their way.

Here was a man overflowing with questions; a man of ruthless objectivity who respected no obstacle in his search for answers. I found myself drawn into a story of madcap adventure in which outrageous anecdote alternated with sober science, showmanship with philanthropy, liberal open-mindedness with a furious religious orthodoxy. In all these ways it was a parable of its time, the hinge between the age of inherited certainties and the age of scientific wonder. In an odd and perverse way, the size of the personality obscured the size of the inner man. Serious scientists are not supposed to be flamboyant. Only now, nearly 140 years after his death, is it possible to recognise the epic scale of Frank's achievements as a pioneer of conservation and one of the all-time greats of natural history.

Tracking his life was a challenge, though the difficulty lay not in eking out scarce material but rather in coping with excess. What to include? What to leave out? His two previous biographers, George C. Bompas (*Life of Frank Buckland*, 1885) and G. H. O. Burgess (*The Curious World of Frank Buckland*, 1967), both had their own ideas of what mattered. There is a telling difference between the two, particularly in terms of what each one chose to omit. Burgess, who aimed to recalibrate Frank's reputation as a scientist, tended to tiptoe around his more macabre experiments and bizarre enthusiasms. Bompas, who was Buckland's brother-in-law, wanted to celebrate him as a man of warmth and honour as well as a pioneer of natural science, which meant omitting some of the stranger aspects of Frank's private life. Along with Charles Darwin and the parson-naturalist John George Wood, Frank was also portrayed in John Upton's *Three Great Naturalists* (1910), though Upton had little of any consequence to add to Bompas.

By far the best authority on Frank's life is Frank himself. He kept two bulging scrapbooks which, in their rich accumulation of letters, posters, cuttings, lists and certificates, provide a running commentary on his entire life from early schooldays to near death. This is a biographer's dream, yet it is nothing to set against Frank's own published writing. There was very little in his life that was not shared with the vast readership of his books and articles, which he wrote largely on train journeys. He did not aspire to a high literary style, but rather to a conversational and artless prose that educated his public even as they laughed at his jokes. For me, he has been a warm, stimulating and occasionally shocking companion whose voice I find difficult to get out of my head. It will certainly reverberate throughout the pages that follow – one point at least on which *The Man Who Ate the Zoo* should require no apology.

Richard Girling, Norfolk, 2016

CHAPTER ONE

The Kit of Parts

The beginning could hardly have been more propitious. 'I am told,' Frank wrote, 'that soon after I was hatched out my father and godfather, the late Sir Francis Chantrey, weighed me in the kitchen against a leg of mutton, and that I was heavier than the joint provided for the family dinner.' A family-sized leg of mutton weighs eight pounds or more – a good size for a newborn. The adult Frank, too, though of modest height, was a giant, vast in girth, immense in his achievements, a huge personality that would fill whatever space it was unleashed upon.

Frank idolised two men in his life, both of whom left indelible marks. The first was his father; the other we will come to later. Father, like son, was a man committed to an ideal. Like his son, he kept company with famous men and original thinkers. But he differed from him in one important way. Unlike the forgotten Frank, Dr William Buckland enjoyed a reputation that would stretch across the next two centuries. He deservedly has a biography of his own – *The Life and Correspondence of William Buckland, D.D., F.R.S.*, by his daughter, Frank's sister Elizabeth Oke Gordon – and he is well represented in scientific libraries. Mrs Gordon's tribute to her father, published in 1894, is well worth seeking out. To know Frank, one first has to know William.

He was a clever boy from a conventional West Country background. His father Charles was rector of parishes in Devon and Dorset, who could pull enough strings to swing his elder son a place at Winchester College and thence, in 1801, to Corpus Christi College, Oxford. A fellowship and holy orders followed in 1809. William's trajectory thereafter was steep. Like many churchmen of the nineteenth century he was as passionate about nature as he was about its Creator, and there was an apostolic zeal in the way he threw himself into the new science of geology, as if, through unpicking His works, he would know the mind of God. His studies took him so far and wide and deep, burrowing beneath the skin of so many counties, that he must have spent more time on horseback than he did in bed. Few men can have acquired a more intimate knowledge of England's hidden anatomy. In a story told by Mrs Gordon, he was heading for London with a friend one night when they lost their way. Dr Buckland dismounted, sniffed a handful of earth and pronounced with the authority of an olfactory satnav: 'Uxbridge!'

His talents did not go unnoticed. By 1818 he was a Fellow of the Royal Society, and a year later Oxford University's first Professor of Geology. His inaugural address, *Vindiciae Geologicae; or the Connexion of Geology with Religion explained*, was published afterwards as a pamphlet. In it he sought to prove, as Mrs Gordon put it, that 'there could be no opposition between the works and the word of God; and that the facts developed by it were consistent with the accounts of the Creation and the Deluge as recorded in the Book of Genesis'. Well, heigh-ho! Science was inviting Genesis to dance, and it wanted to call the tune. William argued that 'the beginning' was not a six-day event but a strung-out process in deep time, spanning the millennia between the origin of Planet Earth and the emergence, through a catalogue of extinctions and mini-creations, of life as we know it. All part of God's grand design. As William explained it in *Vindiciae Geologicae*:

> [A] universal deluge at no very remote period is proved on grounds so decisive and incontrovertible, that, had we never heard of such an event from Scripture . . . Geology itself must have called in the assistance of some such catastrophe, to explain the phenomena of diluvian action which are universally presented to us, and which are unintelligible without recourse to a deluge exerting its ravages at a period not more ancient than that announced in the Book of Genesis.

It was a slick and sophisticated argument, supported by evidence that accounted for the earth's extreme old age and its layered history. But slick and sophisticated are the lodestones of heresy. Oxford in 1820 was still forty years away from the scandalising sensation of Darwin's *On the Origin of Species*. It was stuffed to the pinnacles with reverend gentlemen who cherished their ancient certainties, and for whom 'theory' and 'blasphemy' were cause and effect. Membership of the Church of England at that time was a prerequisite for an Oxford degree – no God, no BA – and academicians had to be ordained. For all William's justifications, the elders did not like the look of geology: it carried a reek of sulphur, a diabolical trimming of eternal truths. According to Mrs Gordon, these were men 'who feared that the study of God's earth would shake the foundations of Christianity'. As late as 1852, when William ventured abroad, the pious Thomas Gaisford, Dean of Christ Church, thought fit to declare: 'Buckland has gone to Italy, and we shall hear no more, thank God, of geology!'

Even so, in a sense, William Buckland had it easy. His important book, *Reliquiae Diluvianae*, published in 1823, sealed his reputation as a visionary. A year earlier he had received the Royal Society's Copley Medal for his work at Kirkdale cave in the Vale of Pickering, twenty-five miles from York, which he had identified as a pre-diluvian hyena's den. Here he had found the remains of twenty-three other species including tiger, bear, wolf, elephant, rhinoceros and hippopotamus, a vivid snapshot of life before the Flood. The evidence for a traumatic deluge, he believed, was there for all to see – in the jumble of soils and rocks over-laying the bedrock, and in the carving of valleys by apocalyptic surges of water. Science, however, expects its proponents to keep up with new evidence. They must be willing to change their minds, and William duly did so. In his best-known work, the *Bridgewater Treatise* of 1836, he accepted that there was no convincing evidence for a universal flood, and became instead a powerful advocate of glaciation – the more plausible idea that the true architect of topography was not water but ice. He burnished his reputation in other ways: by providing the first written description of a fossil dinosaur (*Megalosaurus*, 'the great lizard' discovered in 1824 at Stonesfield, Oxfordshire); by identifying the stones found inside animal skeletons as fossilised faeces, and by giving them the name they still bear – 'coprolites'. To him, a liassic animal turd was a testament in stone, revealing not just a few ancient, undigested fishbones but an entire philosophy of life. For him it proved the 'general law of Nature, which bids all to eat and be eaten in their turn'; carnivora throughout the epochs 'fulfilling their destined office – to check excess in the progress of life, and maintain the balance of creation'.

I said that William had it easy, and compared to Frank he did. Before Darwin stirred up the hornets, William plausibly could assert that Creation was the hand in science's glove. For him there was no conflict. But history would not be so kind to his God-fearing son. Post-Darwin, Frank would be as horrified by all the monkey talk as was his friend 'Soapy Sam' Wilberforce, the reactionary Bishop of Oxford who opposed 'Darwin's bull-dog', Thomas Henry Huxley, in the famous debate at the Oxford University Museum of Natural History in 1860. Darwinism was a hurricane that would leave Frank profoundly disturbed, cling-ing to the pillars of the established Church, torn but obstinate in his faith, the rational scientist casting his lot with Genesis. William died, out of his mind, in 1856, so we cannot know how the author of *Vindiciae Geologicae* might have responded to Darwin's theory of evolution, or how Frank might have squared his position with his father's. The greatest irony is that William Buckland's interpretation of fossils would contribute signifi-cantly to the Darwinian eruption. Geology, wrote the geologist and archaeologist Sir William Boyd Dawkins in his preface to Mrs Gordon's book, 'enabled Darwin to grasp the principle of evolution that now [1894, only fourteen years after Frank's death] influences our view of life as a whole in the same way as the law of gravitation has affected our view of matter, not only in the earth, but also in the universe'. Thus were father and son, however unknowingly, divided by their common interest.

In most other ways Dr Buckland was a yardstick for Frank to measure himself against. William was a charismatic lecturer who liked to entertain his audiences with jokes, funny walks and peculiar exhibits, and who never felt confident until he had raised a laugh. This encouraged some people – sadly includ-ing Darwin himself – to underestimate him. In his autobiography Darwin mentioned him only once, and not until page 102.

All the leading geologists were more or less known by me, at the time when geology was advancing with triumphant steps. I liked most of them, with the exception of Buckland, who though very good-humoured and good natured seemed

Mary Buckland   William Buckland

to me a vulgar and almost coarse man. He was incited more
by a craving for notoriety, which sometimes made him act
like a buffoon, than by a love of science.

Frank, too, liked to spike his lectures with strange and occasionally shocking displays, and he thrived on laughter. He, too,
would be accused of trivialisation, and would be celebrated as
much for the peculiarities of his household as for the importance
of his achievements. Otherwise their paths were markedly different. In 1825, the year before Frank was born, William's rise
through the scientific firmament was matched by his rise through
the Church. He was made a canon of Christ Church Cathedral.
On the last day of the same year he made a fortuitous marriage to
Mary Morland – a highly intelligent woman who was an accom-
plished fossil geologist and marine biologist in her own right,
and who would illustrate many of her husband's books. In his

autobiography *Praeterita*, the critic John Ruskin, a 'gentleman
commoner' at Christ Church from 1837 to 1842, remembers
the Buckland family as 'all sensible and good-natured, with orig-
inality enough in the sense of them to give sap and savour to the
whole college'. Sensible and good-natured they may have been,
and sap and savour they had in abundance, but nobody could
have mistaken the Buckland ménage for a textbook illustration
of Victorian family life.

Frank Buckland was born on 17 December 1826, and raised
in his parents' house in Tom Quad, Christ Church. Altogether
the Bucklands had nine children, of whom Frank was the eldest.
Only five of these, three girls and two boys, survived infancy – a
heavy loss even for Victorian England (infant mortality in the
upper and educated classes then averaged one in five). Frank in
his writing gave much space to his childhood. Some of his memories have a ring of family legend: mythical hand-me-downs. I
find it hard to believe, for example, that the adult Frank could
have had a clear recollection of an event which happened before
his third birthday, though he recounts the story in fine detail. In
November 1829, when he was aged two years and eleven months,
his father bumped into a coachman, 'Black Will', who was 'tug-
ging along in each hand a crocodile about four feet in length',
which he had brought from London to Oxford in the hope of
turning a profit – a matter in which Dr Buckland was only too
happy to oblige. 'Both the crocodiles were put into hot water,'
wrote Frank. 'One died in the water, and the other lived but a
few hours. They were taken over to the anatomy school at Christ
Church, and dissected by the late Dr. Kidd and my father, thinking they would never have such a good chance again, agreed that
they would taste a little bit of the crocodile, and see whether its
flesh was good or not.' A feast ensued, at which the meat was

said to bear comparison with sturgeon or tuna. In another version of what seems to be the same story (though in this case it involved only a single crocodile brought from Southampton), the animal was tipped into Christ Church pond, where young Frank splashed about on its body.

But that was not the end of the affair. In the anatomy school lived an old man called William, whom Frank described as 'the most curious weazen old fellow ever beheld. He wore the old-fashioned knee-breeches, gaiters, and broad-tailed black coat. His face looked like a preparation [by 'preparation' he meant a specimen preserved in spirits], and on his little round head (more like a skull than a head) he wore a very old wig. Altogether he looked much like an injected skeleton with clothes on.' For this apparition the young Frank bore 'the greatest awe and respect'. One can see why. William liked to entertain the little boy with the skeleton of a murderer, which he would let down on a rope just far enough for the toes to rattle on the floor. The 'slightest touch' then 'would make it reel and roll about, swinging its gaunt arms in all directions'. Few boys can ever have had a more powerful inoculation against squeamishness. On the night of the crocodile feast, the old man himself became embroiled in drama. 'In the middle of the night,' wrote Frank, 'there came a furious ringing of the bell, and a messenger from the anatomy school to say that William was dying.' The old man was found sitting up in bed with his wig off, clutching his stomach and 'looking the picture of misery and ugliness'. It turned out that he had taken a large helping of crocodile for his own supper and, in his enthusiasm for free meat, had polished off 'enough for five people'. The consequences were unpleasant but easily cured by what Frank, with untypical coyness, described as 'proper remedies'.

This was not the only story from Frank's formative years that involved the pond. Frank by now was seven. The year was 1834;

the occasion, a visit to Christ Church by the Duke of Wellington. The description is from Frank's diary:

> A live turtle was sent down from London, to be dressed for the banquet in Christ Church Hall. My father tied a long rope round the turtle's fin, and let him have a swim in 'Mercury', the ornamental water in the middle of the Christ Church 'Quad', while I held the string. I recollect, too, that my father made me stand on the back of the turtle while he held me on (I was then a little fellow), and I had a ride for a few yards as it swam round and round the pond. As a treat I was allowed to assist the cook to cut off the turtle's head in the college kitchen. The head, after it was separated, nipped the finger of one of the kitchen boys who was opening the beast's mouth.

Viewed from the twenty-first century, this reads like something from a different world. Turtles are now protected by law, and the green turtle, the kind most often used for soup, is an endangered species. Any parent giving a child a ride on *any* food animal before participating in its slaughter would attract the attention of social services and the RSPCA, if not the police. Yet William Buckland was a deeply caring father who wanted his son to turn out well, not a man who took pleasure in cruelty. As so often in our journey through Frank's life, we must keep a sense of perspective. The nineteenth century cannot be judged by the mores of the twenty-first. Trite but true: we all live in the world we're given and are children of our time. It is not improbable that two centuries hence we, too, will be unkindly judged for our attitudes to children and animals. Frank kept the severed turtle's head, which decades later would become an exhibit in his Museum of Economic Fish Culture at Kensington – a physical manifestation of a state of mind. In effect Frank would parcel up his childhood and carry it around with him until he died, like an internalised how-to-do-it manual. Look! Touch! Take apart! Eat! He had scant regard for books, though he wrote plenty of them himself. From boyhood he was an autodidact who learned through experience, often with little thought for his own safety or for the sensibilities of others. Even in an age which confronted death as often and as intimately as Victorian England did, his approach to mortal remains, whether of man or of beast, was heroically devoid of sentiment. William was the same. Their stomachs could not be turned, nor their resolve deflected. That was William's gift. The lesson he dinned into Frank was that good work required application. 'Never spare yourself,' he said, and it was a lesson that Frank would absorb only too well. Hard work was the drug that would first intoxicate, and then kill him.

The crocodile feast was no rare moment of whimsy. William and Mary's table was famously a post-mortem menagerie of roasted exotics, vermin and superannuated pets. John Upton in *Three Great Naturalists* described a lunch party at which William dished up the pickled tongue of his brother-in-law's horse. The guests 'enjoyed it much, until told what they had eaten'. Ruskin wrote appreciatively of breakfasting at Tom Quad in company with 'all the leading scientific men of the day, from Herschel* downwards', but regretted having missed 'a delicate toast of mice'. The Liberal politician Lord Playfair described an agreeable dinner of hedgehog ('good and tender') but demurred over the crocodile ('an utter failure'). Frank's first biographer, his brother-in-law George Bompas, blandly reported the

---

* The mathematician, astronomer and pioneer of photography John Herschel, son of the great astronomer William Herschel.

consumption of puppies. *And then there was the bear.* William's daughter described the occasion brilliantly:

[William] Buckland was a kind and affectionate father, and always liked to have his children about him. The return from his frequent journeys was awaited by them with eager expectation, for from the famous blue bag would be turned out for them on the dining-room floor some strange (in those days) foreign fruit, such as a bundle of bananas, or a cocoanut in its big outside shell, or a 'forbidden fruit' (lime), which the little ones fondly imagined might have grown in the Garden of Eden. On one occasion, in addition to the blue bag, a large mysterious bundle was brought in, wrapped in a travelling rug. The children were told it was a 'wild beast' of some sort, that it would not hurt them, and that whoever guessed what it was would be rewarded with a penny. The wild beast proved to be the carcases [*sic*] of a bear, which had been seen hanging up outside a barber's shop as an advertisement for the celebrated Bear's Grease – a pomatum for the hair, then much in vogue. The beast had been prepared just like a sheep at the butcher's, only that the skin had been left on the head and the hind-legs to show that it was the veritable animal. A luncheon party was invited to partake of joints of bear, and the fat from the inside was given to the nurse to make pomatum for the family use.

You might suppose it impossible to cap such a story, but that would be to underestimate William's refusal to let anything pass untasted. The English travel writer Augustus Hare (1834–1903) in *The Story of My Life* gives this account of an evening at Scotney Castle, when 'the conversation turned on witchcraft':

Talk of strange relics led to mention of the heart of a French king preserved at Nuneham* in a silver casket. Dr Buckland, whilst looking at it, exclaimed, 'I have eaten many strange things, but have never eaten the heart of a king before,' and, before anyone could hinder him, he had gobbled it up, and the precious relic was lost for ever. Dr Buckland used to say that he had eaten his way straight through the whole animal creation, and that the worst thing was a mole – that was utterly horrible.

According to Frank's later biographer G. H. O. Burgess, William changed his mind about the mole. It was only the *second*-worst thing he had eaten. The worst was a bluebottle. As to the heart, it can't be imagined that William ate the whole organ. A more plausible account is given by the chroniclers of Westminster Abbey, who suggest that the incident occurred in 1848 when Dr Buckland 'was shown a silver locket containing an object resembling pumice stone. He popped the object in his mouth, perhaps to try and find out what mineral it was, and swallowed it. It was in fact part of the mummified heart of Louis XIV of France which had been taken from the royal tomb by a member of the Harcourt family.'

As parents, William and Mary were kind but particular. The children were expected to join in adult conversations at dinner, and to have something interesting to tell their father. Gossip and 'evil-speaking' were not allowed. 'Educated people,' their mother told them, 'always talk of things; it is only in the servants' hall that people talk gossip.' It is a stricture from which the adult Frank seems seldom to have strayed. In his writing, with rare

* Nuneham House, Nuneham Courtenay, Oxfordshire, home of Lord Harcourt, a close friend of William's.

exceptions, a gentle joke at the expense of his detractors was as acerbic as he ever got. Family walks were no mere rambles or 'constitutionals'. The children always had to have an 'errand' – taking cough medicine to a sick bargeman, for example. Sundays were for church and for the weekly treat of a walk in Christ Church meadows or up Headington Hill, led by William for the express purpose of observing trees, plants and stones, and noting the dates when flowers bloomed. It was an idyll of sorts, but a rigorous one.

The house in Tom Quad was a riot of disciplined chaos. The mining engineer Thomas Sopwith noted a staircase made hazardous by a scree of ammonites, fossil trees and bones, and piles of books and papers stacked on tables, chairs, sofas, bookstands 'and no small portion of the floor itself'. Frank's friend William Tuckwell,* in his *Reminiscences of Oxford*, remembered visiting the house to play with Frank and his brother. A side table in the dining room was covered with fossils and with a card bearing the stern admonition 'Paws Off'. Then there was William's menagerie, a seethe of exotic wildlife that used the house as a private jungle and human visitors as fellow denizens – another of William's eccentricities that his son would enthusiastically perpetuate. Tuckwell remembered being served a dinner of mice baked in batter, while 'the guinea-pig under the table inquiringly nibbled at your infantine toes, the bear walked round your chair and rasped your hand with file-like tongue, the jackal's fiendish yell close by came through the open window, the monkey's hairy arm extended itself suddenly over your shoulder to annex your fruit and walnuts'.

---

* The Reverend William Tuckwell, known as 'the radical parson', Christian Socialist, social reformer and headmaster of King's College, Taunton.

Yet none of this was without purpose. Even the mad menus were philanthropically inspired. The ordinary men and women of England were ill-fed and poorly supplied with meat. William's mission was to find new things for them to eat. Mice and puppies were plentiful, so why not? W. C. Williamson, Professor of Botany at the University of Manchester, quoted by Mrs Gordon, applauded William for his dedication to others: 'This craving to be useful in promoting the welfare of the world around him characterised his entire life.' This was no exaggeration. William would not have understood 'blue-sky thinking'. There had to be an end in view, something more than just scratching the itch of curiosity. He saw a connection between geology – understanding the soil – and efficiency in agriculture, putting the soil to work. He liked to say 'there is no waste in nature'. Everything had its uses, even coprolites. This fossilised excrement, phosphates from the bowels of ancient beasts, made excellent fertiliser – a kind of transmillennial recycling scheme. Wilderness offended him. Wild places in the nineteenth century were not cherished as they are now. William's enthusiasm for cultivating morasses, fens and marshes would appal a twenty-first-century conservationist, but the nineteenth century had no idea of the destructive power of its own genius, and nobody would have balanced the needs of birds against the wants of men. You'd have to be soft in the head. The farmlands of England were in depression. Poverty bred discontent which frequently expressed itself in riots. Ricks were burned, machinery broken, butchers and bakers ransacked.

In the field, as in the university, William's progressive thinking was about as welcome as an alligator at high table. Half a century on, Mrs Gordon was still indignant on her father's behalf: 'Agricultural prospects were at a very low ebb, and every sort of advice was looked upon with the utmost contempt and scorn by the John-Trot [i.e. boorish, bumpkin] geniuses of farming. The

more ignorant a man is, the more conceited he is; and, in order
to convince both farmer and labourer that science was any good,
it was very important to be able to point to practical proofs of its
benefits.' Then as now (*pace* the controversies over genetically
modified crops) it was no easy matter to prove the benefits of
techniques which people were afraid to try.

But young Frank listened and learned. He listened when
William took an interest in artificially hatching trout, and listened when he began to think about the migratory habits of
salmon and how fish moved through water. The Reverend
Gilbert Heathcote, sub-warden of Winchester College, told a
story about William that could have been a model for Frank himself. It was William's scientific instincts, not his salivary glands,
that were stimulated when a magnificent turbot was served at a
New College dinner. His mind was on dissection, not carving.
Before the meal could be dished up, he raced ahead of the waiters and surgically removed the turbot's head and shoulders. 'Just
what I wanted,' he said.

William also bequeathed a sharp-eyed scepticism and a loathing of what the modern world would call bullshit. This could be
frightening. Young William Tuckwell confessed that he 'did not
understand his sharp, quick voice and peremptory manner, and
preferred the company of his kind, charming, highly cultured
wife. Others found him alarming; dishonesty and quackery of all
kinds fled from that keen, all-knowing vision.'

Saints in particular had a hard time of it. During the year-long
geological tour of Europe that served as his honeymoon, William
caused outrage in Palermo by denouncing the holy relic of its
patron saint, Rosalia, as the bones of a goat. From Frank himself,
Tuckwell learned the story of the Bucklands' visit to 'a foreign
cathedral, where was exhibited a martyr's blood, dark spots on
the pavement ever fresh and ineradicable. Dr Buckland dropped

on the pavement and touched the stain with his tongue. "I can tell you what it is; it is bat's urine!" ' Everyone should have such a father.

In other ways, William's intolerance now looks on the bigoted side of quaint. Despite his clever wife, he could not accept that women had any useful contribution to make to intellectual life. In 1832 he agonised in a letter to his fellow geologist Sir Roderick Impey Murchison, setting out his (successful) objection to female representation at a meeting of the British Association. Women, he said, would 'turn the thing into a sort of Albemarle-dilettanti meeting* instead of a serious philosophical union of working men'. Even the leading female scientist of the age, Mary Somerville, was browbeaten into staying away – 'for fear', as Mrs Gordon put it, 'that her presence should encourage less capable representatives of her sex to be present'. Hardly the message a feminist would print on her T-shirt. But we need to remember that this was 1832. Women in Britain would have to wait until 1870 before the law gave them the right to own property, and even longer before they had Oxbridge colleges to call their own (Girton at Cambridge, 1873, and Somerville, named for the same polymathic Mary, at Oxford, 1879). It was not until after the First World War that most women in the developed world would get their first, often incomplete smidgeon of enfranchisement, and not until 1928, after a long and bitter struggle, that British women would achieve the same voting rights as men. Viewed from that perspective, William was not quite the harrumphing dinosaur he now seems. His egalitarian instincts may not have evolved any faster than those of his fellow clerics, but neither did he lag behind them: he was a man of his time.

* A slighting reference to the Royal Institution in Albermarle Street, Mayfair.

How all this worked on Frank is difficult to know. There is little evidence of misogyny in his writing, but neither is there much acknowledgement of the part played by his wife Hannah, who seemed an obscure figure even to his brother-in-law biographer George Bompas, who managed both to misspell her name and to omit the embarrassing truth of their early years together. As we shall see, Hannah had a lot to put up with. The unimaginable reality of life with Francis Trevelyan Buckland was not likely to have been covered by any prenuptial advice from her mother or sisters, if she had any.

Frank's own mother was as powerful an influence as his father. 'Most men of mark', observed George Bompas, 'have had . . . remarkable mothers.' Mary shared her husband's love of geology and was a more than capable assistant to him. Had she been born into a different generation, she might have had an outstanding career of her own. She was also a gifted illustrator, a generous hostess, a wise parent whose firmness left Frank in no doubt of her love, and a woman untouched by prejudice. Mrs Gordon described how Mary 'took great interest in the spiritual and bodily welfare of a settlement of Jews living in . . . a very poor part of Oxford'. She did much to help families who lost their homes in a fire there, and at every Feast of the Passover would accept a gift of 'half a dozen of the large thin wafer-biscuits – the "Passover Bread" – in token of respect and gratitude'. Again it is not possible to know what Frank took from this, but he would go on to develop a deep and lasting fascination with other cultures.

We are beginning to see what Frank Buckland was made of. His father was the kit of parts from which he would assemble himself, if not quite in the same order or in the same proportion. Buckland Mark II would be no mere copy of Mark I, but he could not have existed without his blueprint. Frank's mother, too, played her part. She was a guiding light and a necessary moral stabiliser, but no inhibitor of her wilful son's precocious individuality. As she recorded in her diary:

> At two and a half years of age, he never forgets either pic-
> tures or people he has seen; four months ago, as well as now,
> he would have gone through all the natural history books in
> the Radcliffe Library, without making one error in miscall-
> ing a parrot, a duck, a kingfisher, an owl or a vulture.

Frank at this age was recognisably the same person he would be at fifty: impelled by the evidence of his own experience, sceptical of received wisdom:

> September 1829. – I can get him to learn nothing by rote,
> he will not give himself the trouble to do so; his mind is
> always at work on what he sees, and he is very impatient of
> doing that which is not manifest to his senses.

He was still only three when Mary wrote:

> He is certainly not at all premature; his great excellence
> is in his disposition, and apparently very strong reasoning
> powers, and a most tenacious memory as to facts . . . He
> is always wanting to see everything made, or to know how
> it is done; there is no end to his questions, and he is never
> happy unless he sees the relation between cause and effect.

And not yet four when he reached the point at which this book began. This account of it comes from Bompas:

> About this time a clergyman travelled from Devonshire to
> Oxford, to bring Dr. Buckland some 'very curious fossils'.

When he produced his treasures Dr. Buckland called his son, who was playing in the room. 'Frankie, what are these?' 'They are the vertebrae of an ichthyosaurus,' lisped the child, who could not yet speak plain. The dumbfounded clergyman returned home crestfallen.

The die had been cast. At the age of eight, the prodigy would be sent away to commence one of the most bizarre school careers ever recorded in English history.

CHAPTER TWO

Matron's Cat

Frank was prepared for school by his mother. It might better be said that she prepared him for life. Her thoughts, her language, her piety, were a character-building syrup spooned into him daily. The source of Frank's ironclad Christian faith is no mystery when you read the words of Mary Buckland. Here is the letter she wrote to him on his fifth birthday. The handwriting is almost childishly meticulous – a lesson in clarity – though it becomes cramped as she approaches the bottom of the page: an uncharacteristic miscalculation in a woman who took pride in perfection. It is dated 17 December 1831.

My dear little Boy,

You are this day five years old, and I hope that every year you will grow better and wiser.

I pray to God every day, my dear Child, that you may become a good and a wise man, but you must do all you can to try and make yourself good and wise by striving to cure yourself of your faults – You must leave off being impatient, and above all you must be obedient to your kind parents, who love you so dearly, and who never desire you to do any thing but what is for your good.

God loves obedient, gentle children, but the disobedient he will surely punish.

I hope my dearest Frank will be amongst the good and obedient children whom God will reward in this world and the next.

Your very affectionate mother, Mary Buckland

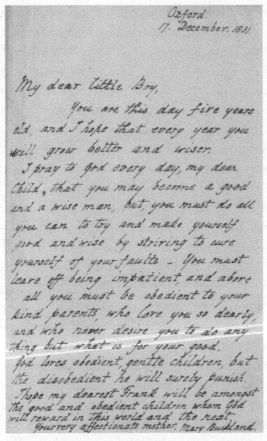

A call to obedience: Mary Buckland's letter, written to Frank at school on his fifth birthday

Some part of this may have been insurance, to keep Frank in good spiritual order in case of premature death. At five, he was not yet beyond the reach of childhood disease. Notes in the family Bible recorded that Frank was 'vaccinated effectually at 2 months old by Mr Bull. Had the measles 1834. Had hooping cough 1835.' Either of these might have carried him off. But this was surely not the limit of Mary's intentions. Perhaps she hoped he would follow his great-grandfather, grandfather and father into the Church. Or perhaps she just wanted him to share her faith, to bind them together *in this world and the next*. On the same day she wrote somewhat less apocalyptically in her diary: 'He reads a great deal to himself such easy books as he can perfectly understand, and he has a happy knack of making himself acquainted with the contents of a book by merely turning over the pages; it seems as if he never forgot anything.'

Frank kept the letter all his life. In fact he seems to have kept almost every shred of paper that was ever addressed to him, or that contained any mention of his name. Many of these are preserved in the grand, classically porticoed building designed by Sir Charles Barry, architect of the House of Commons, for the Royal College of Surgeons in Lincoln's Inn Fields, central London. The college library is on the first floor next to the Hunterian Museum, at the head of a ballustraded staircase hung with portraits of past presidents. Here on an enormous desk beneath pillared galleries and shelves stacked with one of the world's finest collections of surgical literature (some of it reaching back to the fifteenth century), an archivist opens a box and sets down a leather-bound volume the size of a parish Bible. I wait for a while before I open it: set up my laptop, spread out my notebook and pencils (no ink allowed in here), compose myself against the possibility of disappointment. When I fold back the cover it is like a doorway to another world: a treasury of frozen moments. A few of the 259 thick brown pages are empty – their contents perhaps lost or removed before they came into the college's possession – but most are aflutter with stuck-on paper and card, crammed so tight that they overlap. There are letters, posters, William and Mary's marriage certificate, newspaper cuttings, book reviews, school timetables, club membership lists, invitations, menus, cartoons, engravings, photographs, anything and everything, the magpie collection of a man who threw nothing away. A second volume, 308 pages long, is even thicker than the first. At the front of Volume One is Frank's handwritten introduction:

> This Book contains records of my life Beginning 1826 –
> when I first began to put the notes together I was very poor
> I could not afford a better scrap book.
> In Feb 1875 I had the book bound. I thought it better to
> keep the poor old scrap book.
>
> Frank Buckland
> 37 Albany St
> March 1st – 1875

He was forty-eight when he did this, and probably already had some sense of his own mortality: he had only five years left to live. To see and touch these things is oddly affecting, not at all like viewing some dry thing in a museum or even reading Frank's books. These scraps of paper, stained and desiccated though they may be, are more than just witnesses to Frank's life: they are tangible parts of it. To touch them is to touch *him*, and to touch those who mattered to him, most importantly his mother and his father. On his eighth birthday Mary wrote to him again. This time the neat, sloping handwriting looks practised: it is on notepaper so thin that each page bears the mirror image of what is written on the reverse, each word like the ghost of its writer.

My Dearest Frank,

I hope you like the Desk I have bought for you and that you will make good use of it . . . I daresay you will write many a Latin and perhaps a Greek exercise upon it – I hope too you will remember the mother who gave it to you, and who loves you so dearly. Perhaps when you have a very hard lesson and feel inclined to be irritable and out of humour, you may look at your desk, and the thought may come into your heart 'would not my poor mother be vexed to see me so ill tempered', and at this thought perhaps you may cast off the naughty fit, for the sake of the Parent who will never cease to pray that God will send his holy spirit upon you – God bless you my dear child – May you so give an account of the Talents committed to your charge, that like the faithful servant in the Parable, your heavenly master may say to you – 'Well done thou good and faithful servant. Enter thou in the joy of the Lord'.

She concluded by describing the volumes of Thomas Smith's *The Naturalist's Cabinet*, which 'Grandmama' and 'Papa' were sending to him, 'so you will have quite a library of pretty books', and signed herself as before, 'Your affectionate mother, Mary Buckland' – an odd contrast to the informal 'Papa' and 'Grandmama'. This clearly was not a relationship to be taken lightly. Six months later, aged eight and a half, Frank was packed off to his first boarding school, at Cotterstock near Oundle in Northamptonshire, where the headmaster was the vicar, Alexander MacDonald. Not much is known of this period in Frank's life, save that he was acutely homesick. As a sample of his oeuvre, his first letter home was uncharacteristically brief but characteristically misspelled:

> My dear mother will you right to me very often I do not like school so much as i thought I sude be sure and rite very often believ me youre affesenate sone frank buckland
>   It is a very bad letter.
>   be sure and right to me before y. you saile.*

After two years at Cotterstock, Frank moved to a school at Laleham, by the Thames near Staines, run by his uncle, the Reverend John Buckland. For whatever reason – family loyalty, or perhaps just lack of knowledge – George Bompas in his biography of Frank had nothing to say about what went on here. John Buckland does not seem to have been made of the same stuff as his brother. A paper about William written in 1978 by J. M. Edmonds† presents clear evidence of John's early shortcomings. Confusingly, William and John's uncle – their father's brother – was also John. He too was a minister of the Church (vicar of Warborough, Oxfordshire) and did much to smooth William's path into Winchester College. In March 1801 he checked a Latin exercise written by his younger nephew, John, at Blundell's School in Tiverton, which he quickly saw had been 'patched up' from sentences lifted from Cicero and Charles Este's *Carmina quadragesimalia*. Then, as now, there was nothing unusual about schoolboy plagiarism. 'Boys at school are very apt to

---

* William and Mary were about to embark on a tour of France, Belgium and Germany.
† *Patronage and Privilege in Education: A Devon Boy goes to School, 1798*, J. M. Edmonds (Oxford University lecturer in geology, and founder of the Oxford Geology Group, 1957).

practise such impositions in this matter,' wrote Uncle John to
his brother Charles. What made him angry was his nephew's
reaction to being found out.

> With respect to John . . . The truth is, he has been guilty
> of a gross imposition upon me both in the prose and verse
> exercise and further aggravates the offence, by supporting
> it with a wilful lie. I recommend it to you . . . [to] gently
> correct him and remonstrate with him at your discretion
> upon the guilt of lying.

Thirty-six years later, this same John would become Frank's
headmaster. He was by now a man of substance. He had gradu-
ated from Trinity College, Oxford, in 1806 and was made MA in
1809. Among his friends at Oxford was the future headmaster of
Rugby, Thomas Arnold, a reformer whose example would inspire
Baron de Coubertin to found the modern Olympic Games. In
1819 John Buckland entered a partnership with Arnold, who
by now was his brother-in-law (John married Arnold's sister
Frances in 1816), to open a new school at Laleham. Arnold had
long since moved on by the time Frank arrived in 1837, though
this does not seem to have harmed John's reputation. According
to Frank's 1967 biographer, G. H. O. Burgess, one headmaster
of Winchester (I have been unable to discover which one)
exalted John Buckland as 'the father of the English Preparatory
School'. In fact he was not at all the kind of man a headmaster
of Winchester or anyone else would want to brag about. He may
or may not have cured himself of lying, but he had acquired
another vice that quite literally would leave scars on his nephew.
The Reverend John Buckland was a brute.

The most agonising document in the scrapbook is the draft of
a letter from William to his brother. It seems highly significant

that first William and then Frank decided to preserve it: clearly this was no trivial thing to be endured and forgotten. The letter is a howl of outrage. It extends to four pages with some smaller fragments, made almost unintelligible by violent deletions and overwriting, the words not so much imprinted on the paper as slashed into it. Here I must express my gratitude to Burgess, who – a miracle of perseverance – somehow managed to decipher William's scrawl. These are the first few pages as he transcribed them:

Dear John,

I have been very unhappy since my boys' return from school and have postponed from day to day the very painful task which I feel it is my Duty to perform of writing to remonstrate against your mode of punishing children with a round ruler, which is calculated to inflict on their hands and has inflicted on Frank an injury that he will carry to the Grave . . . A portion of the joint has been crushed and the injury is irremediable.

On a former occasion he had a large wound and an extensive gash through the skin of a finger of his R. hand had been cut by the end of the same ruler – and on [Frank's younger brother] Edward's first return a nail had recently been torn off from a finger by the same instrument.

I have put Frank's hand under the care of Mr Tuckwell who . . . [does not] think that Frank's bone will ever recover the injury it sustained. I feel it therefore a Duty to require from you as a condition of my boys' return again to Laleham, an assurance that they shall no more be punished by blows inflicted with a round ruler on the hand more specially on the *Right* hand.

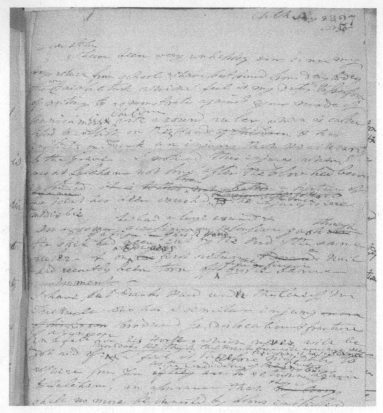

Indecipherable fury: the beginning of William Buckland's letter to his
brother John, raging against his brutal treatment of Frank

But William was no abolitionist. He was careful to make clear
that he was not opposed to physical chastisement in principle.
Indeed, one of Frank's earliest memories was of William forcing
him to spend ten minutes in a gorse bush as punishment for nib-
bling a carriage candle.

It is not clear from descriptions of William's dinners whether
or not he took alcohol, but his normally meticulous syntax soon
disintegrates into the sort of late-night tirade that suggests the
proximity of a bottle. Gone is the controlled precision, the 'sharp,

quick voice and peremptory manner' which had so alarmed
William Tuckwell; in its place a stream-of-consciousness rant.

The surface of the body affords abundant space for punishment by cane or birch which can leave only passing [?] scars
but can inflict no permanent injury let boys be flogged or
caned on the back and shoulders as much as may be needful
but let them not be maimed with an inflexible round ruler or
it may lacerate a tendon and produce lock jaw [tetanus],
or break a bone, or injure for life a joint and at the heart
must render callous and inflexible that skin at the extremities of the fingers in which Nature has placed the most delicate and important Nerves of Touch that occur in the Body.

On he went, page after page, smoking with fury. But there
is a strange omission. His list of approved targets does not
include the all-time floggers' favourite, the buttocks. Even in
the early 1960s, the old-fashioned headmaster of my Home
Counties grammar school could think of no better medicine
than 'six of the best'. Corporal punishment remained legal
in British state education until 1987, and for longer in public
(i.e. private) schools. Most of the developed world, including
Russia and thirty-one US states, has followed the example of
Poland, which banned corporal punishment in 1783, the year
before William Buckland was born. In 1793 the English philosopher John Locke came down firmly on the side of the Poles:
'The usual lazy and short Way by Chastisement and the Rod,
which is the only Instrument of Government that Tutors generally know, or ever think of, is the most unfit of any to be us'd
in Education.'* He argued that children needed some worthier

---

* In *Some Thoughts Concerning Education.*

inspiration, some higher incentive than the avoidance of pain. The 'true Secret of Education', as Locke called it, did not dawn on John Buckland, and it is disappointing to find William – wise though he was in so many other ways – espousing the skill of the torturer: to hurt without leaving a mark. It reminds us what a dark pit the nineteenth century sometimes was, and how long it took to climb out of it. We can't know whether William's letter was actually sent, but Burgess reasonably observes that it must have been – 'otherwise there seems little point in preserving the copy'.

Presumably William did receive the assurance he sought. In any event, Frank stayed at Laleham. According to letters written to his mother ('I have been out evry day this week except Thursday and so have all the boys except one or two and we have had some very good games'), not all the times there were bad, and he seems to have made progress. His father wrote to him in June 1839:

Dear Frank,

I have been much pleased with the amended spelling in your last letter and now that you are going to Winchester presume you will feel it beneath your dignity to retain your former childish and careless habits of neglect as to this matter. I inclose [*sic*] a note to your Uncle desiring him if he think you equal to it to let you read a little of one of the easier plays by Euripides, because they read this author in the Class I think you will be placed in, if you have read ever so little of Euripides. I send a bottle of lemonade powder of which a tea spoon full put into a Tumbler of water will produce effervescence and remain with kind love to Edward

Your affectionate Father

W. Buckland

A month later Frank arrived at Winchester. The coach he travelled in was driven by the very same man who, a few years earlier, had delivered the crocodile Frank had ridden in Christ Church pond. The journey was not without incident. First one bluebottle appeared behind the coach – 'from out of Bagley Wood' – and then another, and another, until there was a swarm. Frank's fellow passengers, mostly other boys heading for Winchester, began to gripe about a bad smell coming from the roof. '[There] was a strong idea among them,' wrote Frank, 'that somehow or other *I* knew from whence this odour proceeded.' They were right. William had given Frank a present for the head-master, Dr George Moberly, a future Bishop of Salisbury, and he had packed it in his trunk. This somewhat challenging gift was a haunch of venison which William had hung for so long that it had become 'over-exalted'. In plain language, it stank – the first of the many olfactory outrages with which Frank would tax his schoolmates' stomachs. They would learn to forgive him, but it would take time. Frank's reward on the coach was to have his head punched. He complained in a letter home that the venison 'stunck very much'. His father's praise for his 'amended spelling' was, and would remain, a victory for optimism over the facts.

Frank seems to have endured without complaint all the ragging and fagging that went with being a new boy at an English public school. William had had good reason to introduce him to Euripides. Classics were the alpha and omega of what Winchester and every other school then taught: Greek and Latin grammar, and rote-learned passages from the Greats. Arithmetic and modern languages were barely acknowledged, and history and geography limited to ancient civilisations. Science got no closer to the curriculum than witchcraft.

Young Buckland's interests lay elsewhere. Bompas provided amusing cameos of him hating cricket, fagging for the 'praefects', taking his turn at making the apple-twig rods which the masters used for beatings, and winning second prize in a 'coffee match'. 'I will now back myself', wrote Frank, 'to make coffee against anybody, even professional cooks.' His recipe:

The coffee must be mixed with cold water, it must then be put on the 'dog-irons' till large bubbles come to the surface; these must be broken one by one with a piece of stick; a piece of the skin of a sole must then be put into the coffee, it must be allowed to boil up again, and then a small cup of cold water thrown in and the coffee allowed to settle.

At night he dreamed of the school holidays. He had bought some ducklings with his pocket money, and couldn't wait to get home to them. His father had promised 'that if I was a good boy at my lessons I should cut their heads off when I came home, on the wood block in the tool-house, with the gardener's hatchet'.

Next to Frank himself, the best guide to the Winchester years is William Tuckwell, whose fond reminiscence, *The Ancient Ways – Winchester Fifty Years Ago*, was published in 1893. Tuckwell has a serious claim to be regarded as his friend's greatest admirer, and he remained so throughout Frank's life. His account is a rolling sequence of amusement, amazement and admiration. This is how he introduces the reader to Frank at Winchester:

Two well-known scientists, Dr. Philip Sclater and Canon Norman,* began I daresay their researches in 'double hedge row' or in the Itchen tributaries. But the naturalist

---

* Philip Lutley Sclater, lawyer, ornithologist, Secretary of the Zoological Society of London, 1860–1902. Probably the Reverend Canon A. M. Norman, an expert on marine invertebrates.

par excellence was Frank Buckland. Short and broad-shouldered, with a shock head of chestnut hair, ruddy cheeks, eyes sparkling with fun – the only boyish feature preserved in the portrait of the man – 'Fat Buckland' or 'Old Buckland', as he was called, was a universal favourite. Every one brightened with amusement at the sound of his loud voice and merry laugh, and the tremendous view halloo which was wont to precede and announce his coming.

Another school friend, Frederick Gale, filled in the detail:

Imagine a short, quick-eyed little boy, with a shock head of reddish brown hair (not much amenable to a hairbrush), a white neckcloth tied like a piece of rope with no particular bow, and his bands sticking out under either ear as fancy pleased him – in fact, a boy utterly indifferent to personal appearance, but good-tempered and eccentric, with a small museum in his sleeve or cupboard, sometimes a snake, or a pet mouse, or a guinea-pig or even a hedgehog.

Frank already was larger than life, his enthusiasm uncontainable, his company infectious. Tuckwell fondly remembered the 'fast and furious' fun of travelling home with him to Oxford, 'the quiet streets of Newbury and Abingdon, the summit of Ilsley Downs, the shades of Bagley Wood, echoing with his jokes, his great post-horn, his chaff of passers by, and the songs he elicited from purple-faced old Stephens the coachman'. Frank's enthusiasms evidently did not extend to the classroom. '[He] was in fact looked upon as "thick" ,' wrote Tuckwell, 'and his compulsory fagging experiences had given him a distaste for games; his delight was to study the habits of live animals and to examine their structure when dead.' Live specimens sharing his locker at various times included a buzzard, an owl and a racoon. Hedgehogs lived beneath a wall in what Tuckwell describes as a *fosse*, presumably some kind of trench, and Frank carried a large white rat around with him until it escaped and joined the resident brown rats behind the wainscoting, thus blessing the school with 'a race of piebalds'. One day, just as the scholars in hall were rising for the master, Frank precipitated 'a good deal of agility' by rushing in with a bottle of ammonia and announcing that his viper had got loose. Typical Frank. Typical pandemonium.

But now we must brace ourselves for some darker deeds. According to Tuckwell, Frank was a 'dexterous taxidermist' who on half-holidays would be found in the boys' washroom, known as Moab, 'plying his scalpel and surrounded by an odour of corrosive sublimate; the subjects being cats, which he snared in their passage through the confectioner's orifice in blue-gate, bats nesting in a hollow plane tree down in Meads, and moles, of whose skins he constructed a very comfortable waistcoat'.

The law these days would prosecute, and Twitter trolls threaten murder to anyone who killed a cat. And yet both Tuckwell and Frank himself could describe his felicidal exploits without fear of opprobrium. Not even Bompas saw any reason to be coy.

The raid on the cats was begun in defence of certain rabbits, kept by some college boys . . . The rabbits had suffered sad depredations at the claws of a cat, and it was determined to defend the weak, and punish the invader, which was speedily wired [snared], and the following afternoon was devoted to the skinning of it. Many other cats after this shared the like fate, including a large black and white tomcat of Dr. Moberly's, brought in with triumph; his skin was a treasure. The fellows in Frank's college chamber were much exercised by a very offensive odour in the room, and in investigation it was traced to the box under his bed, where were discovered the remains of a cat, which he had for some time been in the habit of dissecting member by member on the sly in bed . . .

The skinning of his specimens used to take place in Moab; and the doors of the lockers there, both inside and out, were adorned with the skins of cats, rats, mice, bats, *et hoc genus omne* [and all that sort of thing], duly peppered to prevent smell. His maceration pots at Amen Corner, with heads of hares, rabbits, cats &c., being reduced to skulls, were things to be avoided.

Cats at this time were kept as pets, but a great many were left to roam and fend for themselves. Alfred Roslin Bennett, in *London and Londoners in the Eighteen-Fifties and Sixties,* described hawkers paying as much as 2s 2d for a good cat skin. Ghastly stories were told of cats being skinned alive, 'furs taken in that fashion preserving their lustre longer and commanding a higher market value', and of flayed pets limping home for a last, bloody embrace on their mistresses' laps. Bennett accepted that such stories might have been exaggerated, but recalled that in 1857 a woman was sentenced to three months' hard labour for skinning cats alive. It is not clear from his early writing what Frank thought about animals' experience of pain. This is not an idle question. The father of utilitarianism, Jeremy Bentham, was only seven years dead when Frank arrived at Winchester, and his famous caution on non-human life – 'The question is not "Can they reason?" nor, "Can they talk?" but rather, "Can they suffer?" ' – had not descended quite far enough from the ivory tower to displace the ideas of René Descartes, who had argued that animals had no conscious life and, hence, were incapable of feeling. It would be fair to say that the example offered to young

Frank would have been more Cartesian than Benthamite, but
that the adult Frank would become a forthright opponent of suffering. A letter home, written at Winchester in 1843, shows that
he well understood the quality of mercy:

> Mother (the matron of the sick house) asked me to kill her
> cat on Tuesday, which I did by a blow on the neck; she was
> nearly dead because she had gone away and stayed without
> food for some time, and Mother could not make her eat;
> I skinned her and put her body in chloride of lime, but
> the flesh would not come off as I wanted it, so I cut off her
> skull and boiled it for three hours, and it is now beautifully
> white and clean.

Frank was a persistent scavenger. Tuckwell was with him one
night when they came across 'an enormous dog, drowned many
days before, and swollen by immersion to portentous hideous-
ness'. This they dragged back to school and bribed 'Dungy' the
bell ringer to unlock the belfry door for them to haul the 'exud-
ing carcass' to the top of the tower, where they laid it out on the
leads. Here the dog slowly decomposed, 'his exhalations soaring
upward and afflicting nobody, Frank visiting him from time to
time to see how his symptoms sagashuated,* till on the last morning of the half he "numbered the bones", like Tennyson's Rizpah,
and took them home in a carpet-bag'. Another of Frank's old
schoolmates, the Reverend E. Fox, recalled that the river often
yielded a dead cat or dog, and whenever possible 'the carcase was
rescued and decapitated, the head being carried home to College

---

* Apparently an African American slang term from the Southern States,
meaning 'get along with', or 'endure', popularised by Joel Chandler Harris's
Uncle Remus stories in 1881 (Tuckwell was writing in 1893).

in his handkerchief . . . This was dissected most carefully, each bone by itself – a process which was not very sweet to the olfactory nerves; and then came the question of where to bestow the treasures.' Various locations were mooted. The locker was tried first but failed because of the offence it caused to neighbours; burial failed because cats dug the heads up; immersion in a bowl of water beneath his bed failed because it disgusted the bed-makers, who threw them into the fire. It was then that Frank came up with the wheeze of storing them on the leads. 'Here, exposed to air, sun, and rain, in due season they were bleached, and then, when warranted sweet, they were brought down and ranged in order at the head of his bed.'

The fate of the warden's mastiff was never forgotten by the writer T. W. Erle – 'I remember his dissecting the eye'. Frank included his own account of it in his letter home about the matron's cat.

The warden's large mastiff was shot the other day because it had got the distemper. I got the warden's boy to get me the head: I was very hard at work yesterday cleaning it, i.e. cutting the flesh off; I am going to get it boiled, and then it will be fit to keep: it will make a beautiful skull, I think.

At around this same time, Frank made a decision of the kind usually described as 'fateful'. He resolved to become a surgeon. Typically this was no mere statement of intent. The resolution had to be acted upon, and he began to regard his schoolfellows much as he had looked upon cats and dogs. Tuckwell once again is the best witness.

R–, a commoner with a curiously shaped head, used to relate with a slight shiver that he had overheard Buckland muttering to himself – 'What wouldn't I give for that fellow's skull!' Applying for admission to sick-house on behalf of a patient who had partaken too generously of husky gooseberry fool, he informed the surprised master that the boy had a 'stricture of the colon'. He was wont to offer sixpence to any junior who would allow himself to be bled; and he treated surgically a football-wounded shin, the property of one D–, with such results, that the limb, when shown eventually to a doctor, was pronounced to be in imminent danger of amputation.

A 'Dr Merriman' remembered an actual patient – a man with a bad hand who used to come to the college gate to be treated by Frank, or, in Merriman's words, to be 'experimented upon'. 'In his toys (cupboard),' Merriman wrote, 'he had various bottles and specimens, one very highly treasured possession being a three-legged chicken.' As Frank grew in seniority, so he grew also in audacity and ambition. Never mind dogs, cats and freakish poultry, his dissections now included 'gruesome fragments of humanity' which he bartered from the house surgeon at Winchester Hospital in exchange for eels and trout. It might be thought that his letter home of 1843, giving his account of the matron's cat and the warden's mastiff, was as extraordinary a record as any parent could ever hope or fear to receive of a son's adventures at school. But there was more, and worse. The letter went on:

We had leave out last Tuesday, and I went up to the hospital again, where I met Mr Paul the house-surgeon . . . I walked through some of the wards, and saw the same legs and arms in splints which I saw last week broken and not yet set. I did not see many cases, because I went directly into the operating theatre, where an amputation had been performed that morning: the leg was lying on the table, so I immediately pulled out my pocket-knife and began dissecting it. It had been amputated just above the knee. It was just like a leg of beef with yellow fat, &c, and I daresay it would have disgusted you if you had not been accustomed to dissection. However, I cut away some of the bottom of the foot to look at some muscles, and then I cut off a great piece of skin; I put it in my handkerchief, and took it down to the tanner's to be tanned. The man there turned it about, but could not find out what it was. I shall have it back next Tuesday, I hope tanned like ox-leather. It is very thick, thicker than you would expect. Mr Paul laughed at me for skinning the leg, and particularly when I pocketed the skin. I told the tanner it was the hide of a curious animal (which is true). The boys in College were very much disgusted at my exploits, nevertheless I have had applications made for skin. I shall bring a good piece home with me. Could you manage to send me a lancet by post?

It cannot be wondered at that Frank was so vividly remembered by his contemporaries. Tuckwell promoted him from the 'universal favourite' of *The Ancient Ways* to 'the most popular boy in the school', though his popularity did not extend very far beyond the walls. Frank was a fearless poacher and a thorn in the side of the college's neighbour, the much abused Farmer Bridger. Frank's way with trout was to make a noose of 'the finest pianoforte wire', and – a refined art, this – pass it over the fish's head, much as he used to snare cats. The prefects got wind of this rare skill very early on, and sensibly made him catch their breakfasts. By his own estimation, Frank was also the school's best field-mouse digger. When Farmer Bridger complained about him lighting fires to roast mice in the field, he brought them back to cook in college. 'A roast field mouse', he observed, 'is a splendid *bonne bouche* for a hungry boy; it eats like a lark.' Hedgehog and squirrel pie also appeared on the unofficial school menu, and there were tales of rat-hunts, badger-skinning and rook-trapping. Frank plainly loved the place. Oddly, given his experience at Laleham, he could even look back with fondness at his punishments.

> All I can say . . . is that the jolly good hidings and the severe fagging I got as a lad at Winchester have been of the utmost value to me in after life . . . God bless the dear old place, and all Wykehamists, past, present and future.

I am not inclined to believe that Winchester has ever seen his like again.

CHAPTER THREE

Tiglath-pileser

Having followed his father to Winchester, the obvious next step for Frank was William's alma mater, Corpus Christi, Oxford. Even in the 1840s, however, nepotism had its limits. Frank's talents for autopsy and bonhomie were not enough to compensate for his lukewarm relationship with the Greats. Despite his father's eminence in the university, Oxford did not open its arms to him. Dissected dogs didn't count; what it wanted was Homer. Although William did his best to coach him, Frank failed the entrance exam to Corpus Christi. Magdalen, too, refused him a scholarship. Instead he had to enter Christ Church as a commoner,* though even here he seems rather to have staggered over the threshold. In Frank's scrapbook at the Royal College of Surgeons' library I find this letter from his old headmaster, Dr Moberly, to William Buckland, dated 16 July 1844:

> My Dear Sir,
>   On taking leave of your son Frank from Winchester, I am most anxious to express to you my high sense of his great good conduct and attention while he has been under my care. He

---

* An undergraduate admitted to the university without obtaining a college scholarship or exhibition.

has been unfailingly steady and careful in every thing which he has had to do: and carries away the character of a most amiable and right-minded fellow. – I was very sorry to find that he had made so many mistakes in his examination at Christ Church – sorry, rather than surprised, I must say: for I know how strong his propensity to *blunder* is, even in matters which he knows. But I do trust that when he comes to be known, he will be forced to have more grammatical proficiency.

Frank took up his place in October of that year and was given rooms on the ground floor of the somewhat dilapidated Fell's Building (this would be replaced forty years later by Sir Thomas Deane's Venetian-style Meadow Building). The scrapbook contains a copy of Frank's letter requesting admission to the college, which tradition required him to write in medieval Latin. I am not competent to translate this, but no one can be surprised by Burgess's opinion that the grammar was 'somewhat deplorable'. It would be a while before Frank could take any pleasure in his new situation. Only a month earlier his younger brother Adam had died, apparently suddenly, at the age of nine (that is according to his sister, Mrs Gordon; another account gives his age as six). The boy was buried alongside another brother, Willie, and a sister, Eva, in a vault at Christ Church Cathedral. To recover from their bereavement, William took his family on a geological field trip to the Jurassic coast of west Dorset.

Despite all this, the Frank who arrived at Christ Church was the very same Frank who had departed Winchester. One of his contemporaries there, Herbert Fisher,* first caught sight of him

---

* The historian Herbert William Fisher, author of *Considerations of the Origin of the American War*, who was tutor to the future Edward VII and Keeper of the Privy Seal.

on the Oxford coach. Frank struck him as 'a very strange-looking little fellow . . . but [I] saw at once that he was unlike anyone else'. Fisher was impressed firstly by Frank's unstoppable flow of talk, and then by his impish precocity. When the coach changed horses and one of the replacements turned out to be lame, Frank darted down from the box, lifted the affected foreleg and – 'much to our amusement and that of the coachman and ostlers' – confidently diagnosed the injury. When the coach got under way again, Frank removed his hat and took out a large moth, which 'he began to examine carefully, calling our attention to its characteristics'.

'Recalling that journey,' wrote Fisher, 'I cannot help remarking that Frank was exactly then what he always continued to be. For of all the men whom I have known, I should think that he must have changed the least with advancing years.' Frank's way of growing up was simply to become a bigger boy – 'a child of nature', as Fisher put it, 'with a mind full of child-like mirth and gaiety; yet rendered serious by the eagerness with which he scanned all natural objects, so intense that no room was left for the slightest thought of self. He seemed to assume that everyone must take as much interest in these things as himself, and this imparted that freshness to his conversation which made him so attractive a companion to people of every kind; for he knew no distinction of persons.' Fisher might have been surprised by just how far *people of every kind* eventually might stretch.

At Oxford, according to Fisher, Lord Dufferin ran 'a small debating society'* whose members were expected to submit essays for discussion. Earnest young gentlemen liked to pontificate upon great issues and great men – solid, well-worn subjects

* Frederick Hamilton-Temple-Blackwood, 1st Marquess of Dufferin and Ava, a future Governor General of Canada and Viceroy of India. The 'small debating society' was the Oxford Union, of which he was president.

like Charles I and Oliver Cromwell. Frank's first topic was 'Whether Rooks are Beneficial to the Farmer or not', which he followed with a dissertation upon Egyptian natural history, for which his authority was Herodotus, and a history of the dodo. The young gentlemen 'were of course almost dying with suppressed laughter at this delicious innovation'. Fisher implied that this was some kind of jape on Frank's part, but it sounds more like a straightforward expression of things that interested him. Of course there was laughter. With Frank there always was. But it didn't mean he lacked seriousness. Bompas, writing forty years later, cited yet another fellow student, the cleric and academic Richard St John Tyrwhitt, who put Frank's eccentricity into its rightful context.

> It is not quite satisfactory to look back from the present, when the natural sciences are fully and effectually taught in Oxford, and when earnest study in any of them is sure to meet encouragement and ample reward, to a time when an energetic student of physics and born field-naturalist was considered simply off his head for caring about nature.

Overcoming the opposition to natural science was one of the great causes célèbres of Frank's life. Even now the victory is far from complete. Despite the interest in wildlife encouraged by Sir David Attenborough and his followers, children remain shockingly unaware of how living things relate to each other and why they matter. A public school headmaster told me recently that he wanted ecology to figure more prominently in his school's curriculum, an idea he seemed to think was both new and enlightened. It is a not altogether reassuring illustration of how far Frank Buckland was ahead of his time.

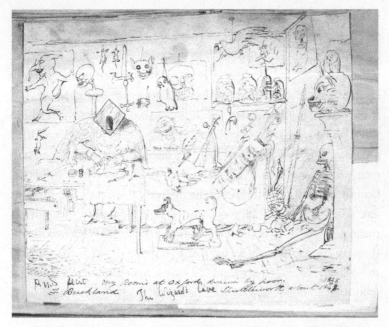

'The Wizard's Cave' – a sketch by Philip Ughtred Shuttleworth of
Frank's room at Christ Church

Tuckwell as ever was the man for detail. He described Frank
hosting 'unique breakfasts' in his rooms, dressed in blue pea
jacket and German student's cap, and 'blowing blasts out of
a tremendous wooden horn'. It was not just the menus that
made these occasions impossible to forget. Frank's free-range
menagerie was, if anything, even more intrusive than his father's.
Tuckwell noted marmots, a dove, monkey, chameleon, snakes,
eagle, jackal and pariah dog. Bompas added guinea pigs, squirrels,
dormice, tortoises and green frogs as well as skeletons, stuffed
specimens and 'anatomical preparations'. There is no record
of a fox, but this did not stop Frank from hollering his *view
halloos*. Unsurprisingly, Bompas and Tuckwell both expended
many words on Frank's bear cub. Despite Byron's example at

Cambridge, an ursine at university was still a rare and alarming sight, guaranteed to attract attention. Ruskin, a passionate admirer of Byron, came over all lofty about this. Frank, he declared, 'was too fond of his bear cub to give attention enough to the training of the cubbish element in himself'. That might be arguable, but there can be no doubt that the animals responded anarchically to the looseness of Frank's control. For the eagle, Oxford provided a pleasantly varied smorgasbord. It killed a hedgehog, chased and nearly caught a terrier, killed and ate (in Frank's own words) 'a beautiful little kitten, the pet of the nursery . . . Several Guinea pigs and sundry hungry cats too paid the debt of nature through his means . . .' At the breakfast table it ravaged a ham before making off through the window with a partridge. It next turned its attention to the college chapel, where it lurked by the door and attacked anyone who tried to enter. The chapel had a particularly hard time of it. As if the eagle were not enough, it had also to put up with the bear. According to Tuckwell, this resourceful animal

found [its] way into the chapel, at the moment when a student was reading the first Lesson, 2 Kings xvi, and had reached the point at which King Ahaz was on his way to meet Tiglath-pileser, King of Assyria, at Damascus. So far as that congregation was concerned, the meeting never came off; the bear made straight for the Lectern, its occupant fled to his place, and the half-uttered name on his lips was transferred to the intruder.

For the rest of its life the bear was Tiglath-pileser, or Tig for short. He had arrived in college at the age of six months, when Frank had fitted him out with academic cap and gown, taken him to drinks parties and boating along the river. Sir Charles Lyell, author of *Principles of Geology* and a close friend of Darwin, described Tig's attendance at a meeting of the British Association at Oxford in 1847. Frank formally presented the bear to Lyell himself, then to the Prince of Canino (Charles Bonaparte, a nephew of Napoleon) and to various other grandees whose reward was to have their fingers seized and vigorously sucked. According to Bompas, Tig was cornered by the poet Richard Monckton Milnes, the future Lord Houghton, who attempted to mesmerise him. 'This made the bear furious, but he gradually yielded to the influence, and at last fell senseless on the ground.'

This was all too much for the dour and humourless Dean of Christ Church, Thomas Gaisford. 'You or that animal, Mr. Buckland, must quit the College.' Tig was rusticated firstly to Islip, a village seven miles from Oxford, where, for historic reasons, Frank's father held the living. A series of catastrophes (terrifying the cook, chasing sheep, sucking unsuspecting fingers) climaxed in a raid on a grocer's shop, with Tig 'devouring the sugar and sweetstuff, and terrifying the shop-woman out of her wits'. In November 1847 he was shipped off to the Zoological Gardens in Regent's Park, where according to Bompas 'he died some time after in an effort to cut his teeth'. We shall next meet him stuffed, guarding a passage in the Westminster Deanery.

There is a temptation – easy to fall into, hard to resist – to depict Frank through a series of ripping yarns in which he appears as an amalgam of Dennis the Menace and Doctors Frankenstein and Dolittle. This is to get him badly wrong. The 'cubbish element', as Ruskin put it, had nothing to do with any weakness of character. Ruskin was confusing style with substance (a dangerous habit for an art critic). Frank's dedication to science, and to the work that went with it, was beyond Stakhanovite: in middle age it would become literally self-destructive. Tyrwhitt confessed to having shared Ruskin's misapprehension. 'I think that the rest of us, who only thought of Greek and Latin reading, if we read at all, never quite understood the reality of the value of the work Frank was engaged in, or that he was in fact educating himself much better than most of us were doing.'

Frank spent the long vacations of 1845 and 1846 at the University of Giessen, under the tutelage of the great German chemist Justus von Liebig.* To make sense of Liebig's lectures he first had to learn German, and did so at a speed that might have impressed even Ruskin. His letters were full of colourful observations about the way the Germans dressed (quaintly) and ate (with 'perfect indifference' to fat or lean). Old habits died hard. A hunter in the forest handed him a dead polecat which, to strip the flesh from the bones, he parked on an ants' nest. He was annoyed when some unseen hand removed it, but was soon diverted by his upstairs neighbour at Giessen, an anatomist whose dog had 'a silver tube in his stomach, which is stopped up with a cork; the object is to get gastric juice fresh for the lecture; the dog seems very well but rather unhappy'. Outside, Frank took a particular interest in Giessen's unusually noisy frogs. 'I have bought some . . . They are half green, their legs being brown as the English frogs; I inflated one and squeezed the air towards his mouth, and out of the sides of his jaws sprang two bladders. I suppose these are used to make the curious loud noise for which they are remarkable.' A dozen tree frogs, kept in a bottle, accompanied him on his journey back to Oxford and angered his sleep-deprived fellow passengers with their racket. In the end, though Frank got them safely home, they fared little better than the polecat or the ants. On his second day back, a 'stupid housemaid'

---

* Founding father of organic chemistry, discoverer of nitrogen, father of the fertiliser industry and, by descent through his Leibig Extract of Meat Company, the Oxo cube.

opened the bottle. One of the frogs 'croaked at that instant, and so frightened her that she dared not put the cover on again. They all got loose in the garden, where I believe the ducks ate them.'

On his next journey home, in the following summer, he took some German red slugs. These too caused anguish. He woke at midnight to find that two of them had escaped and were making tracks across the bald head of a sleeping passenger in the coach. He left them where they were and quietly hopped off at the next stop. The interaction of his specimens with other people was a cause of irritation, merriment and farce throughout his life, though this was a rare instance of him ducking the consequences.

The previous year, 1845, had seen a profound change in the Buckland household. The Dean of Westminster, 'Soapy Sam' Wilberforce, became Bishop of Oxford and, on the recommendation of Sir Robert Peel, William Buckland was invited to replace him. Shortly afterwards William was granted the living of Islip – an ancient piece of patronage which, as Mrs Gordon explained, had been 'bequeathed by Edward the Confessor to the Abbot of Westminster'. Here it was that Tiglath-pileser embarked on his final rampage.

The move delighted everyone. Peel declared that he had 'never advised an appointment of which I was more proud' – an opinion which Mary Buckland warmly endorsed in a letter to Sir Philip Egerton.*

I think Sir R. Peel has shown much moral courage in making choice of a person of science, for it was sure to raise a clamour, and among good people too. It has always been quite unintelligible to me how it happens that on the Continent,

---

* Sir Philip de Malpas Grey Egerton, palaeontologist, expert on fossil fish and Conservative Member of Parliament.

where there is far less religion than in England, a man who cultivates Natural History, who studies only the works of his Maker, is highly considered and raised by common consent to posts of honour . . . while, on the contrary, in England, a man who pursues science to a religious end . . . is looked upon with suspicion, and, by the greatest number of those who study only the works of man, with contempt. Perhaps you can comprehend this anomaly, I cannot.

It is true that there had been scant respite from the hostility which science in general and geology in particular aroused in the Oxford fundamentalists. A few years earlier Baron von Bunsen, soon to be Prussian Ambassador to the Court of St James's, had observed in a letter to his wife: 'Buckland is persecuted by bigots for having asserted that among the fossils there may be pre-Adamite species. "How!" say they; "is that not direct, open infidelity? Did not death come into the world by Adam's sin?" I suppose then that the lions known to Adam were originally destined to roar throughout eternity!' A hundred and seventy-five years later, anti-scientism still flourishes. Many people of faith, like William himself, have managed to accommodate a changed view of life's origins without letting go of their beliefs, but there are many within the Islamic, Judaic and Christian traditions who abhor the theory of evolution, and many more for whom science is the antagonist of nature and, hence, of its creator – not quite a conspiracy of infidels, but almost. William recalled being invited by a rector in the West Country to agree with him that the great geologist William Smith, author of the first nationwide geological map, was *an ignorant old humbug*. Another told him: "'Tis very well for you to humbug those fellows at Oxford with such nonsense, but we know better at Mugbury!"* And so it goes on.

* Probably a misspelling of Musbury, Devon.

Wherever natural or medical science is at the cutting edge – in gene therapy, for example – there you will find religious conservatives erecting barricades. Mary Buckland's *anomaly* may have changed its polarity – in most advanced societies now it is the fundamentalist, not the scientist, who commits the thought crime – but for all that the two strands may intertwine, they cannot be fused together. This was the difficulty that pursued William into Westminster, and that would create such a dilemma for Frank.

It is a rare father who is a hero to his son, but William was idolised by Frank both during his lifetime and afterwards. Even in death he was an inspiration, the superlative to which his son always aspired. Academically Frank never quite made it. His destiny lay far beyond the margins of academic teaching. He didn't want to learn: he wanted to *find out*. There was never any professorship on his horizon, and his exam successes seem to have been more of the 'bare scrape' than 'flying colours' variety. His college life after William's move to London continued much as before. The troublesome eagle found a new home in a courtyard next to Westminster Abbey, but it made little difference to Frank's propensity for annoying the elders. The security arrangements for his snakes were so lax that the mellow tranquillity of Christ Church was frequently shattered by stampeding undergraduates and college servants. The theologian Henry Liddon, quoted by John Upton in *Three Great Naturalists*, saw Frank as a kind of grubby precursor to *Just William*.

One day I met Frank just outside Tom Quad. His trousers' pockets were swollen out to an enormous size; they were full of slow-worms in damp moss. Frank explained to me that this combination of warmth and moisture was good for the slow-worms and that they enjoyed it. They certainly were very lively, poking their heads out incessantly, while he repressed them with the palms of his hands ... The jackal was, I might almost say, a personal friend. He was fastened up outside Fell's Buildings; and I recollect, under some odd and painful irritation, he used to go round and round, eating off his tail. Frank expressed great sympathy with him, modified by strong curiosity – he wondered how far Jacky would eat up into his back!

What Jacky did eat was four or five of Frank's guinea pigs, which he cornered beneath the sofa. Frank meanwhile continued to satisfy his own epicurean curiosity. One of his friends was Edward Cross, proprietor of the Surrey Zoological Gardens at Kennington, who one day told him his panther had died. This was too good a chance to pass up. 'I wrote up at once to tell him to send me down some chops. It had, however, been buried a couple of days, but I got them to dig it up and send me some. *It was not very good.*'

In October 1847 Frank failed his BA. We can't be sure how much of a surprise this was, but we can see from the scrapbook how he felt when he passed at the second attempt in May of the following year. He had made his own hand-ruled calendar for the months of February, March, April and May, ticking off each day in the countdown to 12 May, where he added the portentous and possibly dread words: 'Public Examinations commence'. The examinations were public indeed. Such was Frank's popularity that, according to Henry Liddon, 'almost the whole undergraduate world of Christ Church' turned up to watch his viva voce on 15 May. On this day Frank recorded in exuberant scrawl: 'Hurrah!! FTB through. *Labor omnia vincit.*'

Tucked between the scrapbook's fluttering pages is a packet containing two small lozenges of once-white gauzy material on a thin white ribbon – the formal neck-bands that Frank wore for his examination. They rest weightlessly in my palm, and I wonder if he was still wearing them when he inscribed his *Hurrah!!* For him they were precious mementoes of an important day; for me they are an intimate hands-across-the-centuries connection with a man whose mind I am slowly beginning to understand. He might not have been squeamish but he was not empty of feeling. Frank Buckland was a deeply sentimental man, and it was sentiment that propelled him. But it did not keep him in Oxford. He pocketed his degree, rounded up his menagerie and headed for London.

The Westminster Deanery had plenty of room for them all, animals and people alike. Mary Buckland did not record the number of rooms – perhaps she never totted them all up – but she did count the staircases. There were sixteen. It did not, she felt, 'look like a very lively abode, for it opens into the Abbey and contains the Jerusalem Chamber'. This was, and is, one of the most historic rooms in England. It was in front of the fire here that Henry IV died in 1413; here that the authors of the King James Bible convened in 1611; here that Sir Isaac Newton was laid out in 1727 before being buried in the Abbey in the presence of Voltaire. Here, too, very likely, that Frank's menagerie wandered, sniffed and demonstrated its lack of house-training. The Bucklands did not have the house all to themselves. Mrs Gordon remembered 'the excellent portress Mrs Burrows' who worked before the fire in the Robing Room 'to air the linen surplices of the canons in residence, as it was highly necessary that these elderly dignitaries should be protected as far as possible from the well-known deadly cold of the Abbey'.

Rats ran everywhere, and the upper rooms were so cold that the servants were unwilling to sleep in them. They complained

of 'queer noises' and of strange gusts, like the breath of polter-
geists extinguishing their candles. Terror reigned when a length
of wainscoting fell down one night, opening up a deep hole like a
well. Frank and his younger brother Edward seized the chance
for adventure. They let down first a lighted candle to check the
air, and then a rope which they descended to find a crumbling
and worm-eaten bedstead and table. According to Mrs Gordon,
this had been the hiding place of Francis Atterbury, Dean of
Westminster from 1713 to 1723. He had needed it because
of his support for the Jacobite Pretender, but it did not save
him from arrest. He was marched off to the Tower of London in
1722, and exiled to Paris, where he died in 1732.

No period of Frank's life would leave a deeper impression.
William Buckland was a gregarious man for whom the exchange
of ideas was as essential as air and water. He kept open house
at the Deanery, where a never-ending stream of visitors braved
the hazards of breakfast, lunch and dinner. Conversation would
not have lingered long on the weather or even on the hedgehog, tortoise, potted ostrich, rat, frog and snail that now graced
the menus ('Tripe for dinner,' reported one guest, 'don't like
crocodile for breakfast'). I doubt that even affairs of the Church
would have detained them for long. As Mrs Gordon modestly
put it: 'The house was the centre . . . to which men of science
resorted, and where many of their discoveries were explained or
illustrated.' By *men of science* she did not mean the occasional
jobbing apothecary. Guests recorded in Frank's diary included
Sir Humphry Davy, Sir John Herschel, Thomas Henry Huxley,
the geologist Sir Roderick Impey Murchison, Sir Charles Lyell,
Richard Owen, Joseph Hooker, Michael Faraday, William
Whewell (the polymathic Master of Trinity College, Cambridge,
originator of the terms *scientist* and *physicist*), Isambard Kingdom
Brunel, Robert Stephenson, Baron von Bunsen and the Swiss

palaeontologist Louis Agassiz. From the worlds of politics, theology and the arts came the Duke of Wellington, the social reformer and former Lord Chancellor Lord Brougham, Lord Rothschild, Sir Robert Peel, the Archbishop of Dublin Richard Whateley, the poet Samuel Rogers and John Ruskin. 'If a man is known by the company he keeps,' observed John Upton, 'certainly Frank Buckland in those days was entitled to be regarded as a man of cultivated tastes and high attainments.' This might be pitching it a bit high for a young man yet to make his mark, but his grooming was impeccable.

William's influence was important in another way too. His ideas were not just learned abstractions, for what was the use of theory without practice? The boys of Westminster School were particular beneficiaries of his zest for reform. Before William, this ancient institution within the Abbey precincts had fallen into disrepair and disrepute. 'In that foundation,' wrote Sir Roderick Murchison,

> education could be no longer obtained except at costly charges, and even when these were paid, the youths were ill fed and worse lodged . . . All these defects were speedily rectified by the vigour and perseverance of Dean Buckland. The charges were reduced; good diet was provided; the rooms were well ventilated, and the buildings properly under-drained; so that, these physical ameliorations accompanying a really sound and good system of tuition, the fame and credit of this venerable seminary was soon restored.

This bland encomium disguised one of the great cornerstones of William Buckland's character: the strength of his convictions, and his grit in the face of adversity. Sanitation in London during the 1840s was little better than medieval, and had been kept that way by men with a medieval mindset. Only a very few forward-thinkers, William among them, had begun to see the connection between sanitation and health. It would not be until 1854 that John Snow would conclusively disprove the prevailing *miasma* theory of disease transmission and introduce the world to *germs*. The school's lavatories drained into a ditch filled with what Mrs Gordon euphemistically described as 'black mud' – evidently a creek of the Thames which 'came up as far as these buildings; but apparently no tide ever succeeded in washing back to the river any of its murky contents'. William needed all his determination to face down the sceptics. One Westminster schoolmaster, the Reverend E. Marshall, doubted that 'any one with a less commanding scientific reputation than Dr. Buckland, even with all the power of the Dean, could have overcome the prejudice which at that time was entertained against the alterations'. William proved his point the hard way. In May 1848 workmen broke into some old drains in Little Dean's Yard, which resulted in a number of people, William and two of his daughters among them, contracting typhoid (also called 'Westminster fever' as the outbreak was confined to the Abbey precincts). Unlike some of the other, more unfortunate sufferers, William and the girls recovered their health, and Mrs Gordon reported that he 'lost no time in applying his scientific knowledge to the thorough cleansing and making of sewers. The system of pipe-drainage which he introduced was the first of its kind ever laid down in London.' Four hundred tons of 'foul matter' were removed during the clean-up, but this did not stop the more idiotic of his opponents claiming it was William's own reforms that had caused the outbreak. This was one example among many of William Buckland's commitment to change, and his faith in science as a force for public good. On 15 November 1849 he preached a sermon in the Abbey which would do credit to any modern socialist firebrand.

> The greater number of the poor who perish are the victims of the avarice and neglect of small landlords and owners of the filthy, ill-ventilated habitations in which the poorest and most ill-fed and helpless are compelled to dwell. Fatal diseases are continually engendered from lack of adequate supplies of water, withholden from the dwellings of the poor by the negligence of the owners, or by the jealousy of interference by public officers or public Boards of Health . . . It will be the fault of man, and of the selfishness, or the folly, or avarice of the owners of poor houses, or of the jealousy or pride of officers and interested individuals, and it will be the fault of Parliament also, if we do not instantly begin to remedy these crying evils.

We need look no further for the root of Frank's idealism. Burgess believed that his reason for taking up surgery was to please his father. This may be so – it surely did please him – but I am certain that Frank wanted to *emulate* William as much as he wanted to make him happy. What could be more William-like than this entry from Frank's diary, written in 1848? 'My object in studying medicine (and may God prosper it!) is not to gain a name, money, and high practice, but to do good to my fellow-creatures, and assist them in the hour of need.' The reasons were all Frank's own. In May of that same year, only a few days after taking his degree, he began his surgical training at St George's Hospital.

For Frank, the Deanery offered limitless opportunities for adventure. Even before he took up residence there, his exiled sea eagle had made its presence felt. The various accounts of the day, including Frank's own, muddle the detail. My favourite is Mrs Gordon's, so I'll begin with her. The 10th of April was the date marked for the event that was supposed to transform the fortunes of the British working class – the great Chartist demonstration at Kennington, opposite the Houses of Parliament in south London. Violence was anticipated, and 'every preparation was made to secure the Abbey and its precincts from any rough treatment by the mob'. In the event, the demonstration was a damp squib that fizzled out in the rain, effectively destroying both the movement itself and the reputation of its leader Feargus O'Connor. Earlier in the day the omens had looked propitious. 'As Feargus O'Connor was earnestly addressing the petitioners . . . an eagle was seen to be soaring over their heads and flying towards Westminster! This naturally was hailed as an excellent augury!' The eagle was Frank's, enjoying what would turn out to be its last hurrah on what in many ways was a typical day for both of them. Frank later confessed that he often had fights with this bird, which 'was of a rather savage disposition', and described what happened after one of its misdemeanours at Oxford:

One day I had a row with the eagle about something or other – I think it was about a dead cat – and, pouncing down from the bough where he was kept, the rascal fastened both his claws into the front part of my right thigh, causing the blood to flow copiously, and, at the moment, giving intense pain; the claws had to be taken out one by one, beginning with the hind claw, and I ordered a broomstick to be put within the grasp of the bird, and he clutched hold of the broomstick with the tenacity of a vice. I did not want to spoil my bird by cutting his claws, but nevertheless I punished him by putting wine corks on to his talons, and I made him do penance in them for a week.

It was through experiences like this that Frank was able to represent himself as an expert on how to treat dangerous birds.

It is very easy to handle an eagle . . . if only you know how. The brute always strikes first with his claws, and then pecks with his bill. Remembering this, – allow the eagle, hawk, or owl, to clutch something or other, say a broom-handle or a walking stick; then quickly throw a Scotch plaid, or a blanket, over his head, when you may release the stick, let him clutch a bit of the plaid, tie his legs, and he can be carried anywhere.

It was indeed a plaid that settled the issue on 10 April. After perching on a high pinnacle of the Abbey and attracting a throng of excited onlookers,* the eagle was lured back to its courtyard by a sacrificial chicken tied to a stick. 'Whilst he was busily engaged in devouring the chicken, a plaid was thrown over his head, and he was easily secured,' explained Frank. The bird's reward was to follow Tiglath-pileser into exile at the Zoological Gardens in Regent's Park. Tiglath-pileser himself had now been stuffed and, with Billy the Hyena, haunted the passage between the Deanery and the Abbey. Whatever consternation these two aroused, it was nothing compared to the panic caused by Frank's loitering snakes. There was also the pet monkey Jacko, and a female ape, Jenny, brought from the Rock of Gibraltar, which, said Bompas,

---

* An escaped eagle would have caused a sensation. Even in 1965, when a male golden eagle, Goldie, spent twelve days on the loose from London Zoo, it dominated newspaper headlines and television bulletins across the world, and caused severe traffic jams around Regent's Park.

'used to lift up her hideous face and kiss her master with great profession of affection'. It can't be imagined that many kisses were forthcoming from the young woman who sat down at the Deanery piano to play quadrilles, with the reasonable expectation of enjoying an evening of music and dancing. Telling her to be a 'good girl' and not make a fuss, Frank coiled a snake around her neck, and one round each arm, while she went on playing. His assurance that he had extracted their fangs might have soothed her mortal fear, though it might not entirely have restored her pleasure. Frank at this time failed to understand that his enthusiasms were not shared by all, and the best that could be hoped for at the Deanery was that people would adjust to his ways. 'His sisters', wrote Bompas, 'were so often bedecked with similar reptilian necklaces and armlets, that they became used to the somewhat clammy, crawling sensation, which is a drawback to such ornaments.' Frank also kept a collection of fifty or sixty rats, and these too were (Bompas again) 'brought up at evening parties for the amusement or torment of the visitors'. The rest of the menagerie included hedgehogs, tortoises, cats, bats, hawks, owls, 'an aviary of various birds', lizards, goldfish and newts. But it wasn't all about alarms and nuisances. The Deanery could also present Frank with uniquely Frank-like ways of making himself useful. When a cat expired in the diapason pipe of the rebuilt organ in the Abbey, he was able to call upon his angling skills and fish it out with a salmon hook.

St George's Hospital, now removed to Tooting, is still one of Britain's biggest teaching hospitals. When Frank began his surgical training in May 1848, it had not long moved from Chapel Street to Lanesborough House at Hyde Park Corner, a grand design by William Wilkins, architect of the National Gallery, University College London and many of the nineteenth-century buildings of Cambridge University. It is now the Lanesborough, reputedly London's most luxurious hotel. It's hard to know who would be more surprised by the contrast – the patients who suffered and died in its 350 beds in the nineteenth century, or the royal personages, political grandees and celebrities who luxuriate in its suites in the twenty-first. Here it was, having sworn to his diary that he would 'be a great high priest of nature, and a great benefactor of mankind', that Frank embarked upon the dangerous, hit-or-miss enterprise that was nineteenth-century surgery.

CHAPTER FOUR

Poor Little Physie

Eighteen forty-eight was the year of revolutions. The disgruntled populaces of France, the Austrian Empire, the German and Italian states, Denmark, Hungary, Poland and many others across Europe and Latin America struggled with little success to throw off their oppressors. The best Britain could manage was the Chartist movement, which perished under the eye of Frank's eagle in April. But if it was a bad year for revolutionaries, it was an auspicious one for surgeons. Two and a half miles by road from St George's Hospital – in Gower Street, Bloomsbury – stood the red-brick colossus of University College Hospital. It was here, just a few months after Frank was admitted to St George's, that the young Joseph Lister first took up his knives. This is more significant than it looks. It meant that Lister's great contribution to surgery, antisepsis, was still nineteen years in the future, and that surgeons still laboured, and their patients suffered, without it. Frank himself, who seems to have been immune to horror, left no very detailed account of the nineteenth-century operating theatre. I rely instead on Lister's biographer, his nephew Sir Rickman John Godlee,* for

---

* In *Lord Lister*, 1917. Godlee was also a surgeon, one of the first successfully to remove a brain tumour.

an idea of what these charnel houses were like. There had been
progress of a kind. Anaesthetics were a recent discovery, but
their introduction was slow and the sawbones carried on their
chancy trade much as they always had done.

In spite of the fact that there was no longer the same
urgent need for haste, operations were still performed
with breathless speed; indeed the idea that the more
quickly the process could be completed the better for
the patient, lingered on for another quarter of a century.
Students in the fifties, therefore, saw precisely the surgery of the pre-anaesthetic days, robbed only of its most
shocking feature – the pain inflicted by every operation.
There was a plentiful display of manual dexterity; but,
looked at as a whole, it was rough and ready surgery.
At University College Hospital one theatre of modest
dimensions, containing one small instrument cupboard,
a sturdy wooden table, a single gas jet and a solitary washing basin, served for every purpose, including the novel
one of administering the anaesthetic . . . [Lister] knew it
was a good sample of London surgery, and had no reason
to criticize it. But, when he followed up his cases in the
wards, there was much to make him think, much to make
surgery a sad calling for a beginner. It seemed to be a lottery whether patients recovered or died . . . He could not
understand how his teachers had the hardihood to use the
knife except under the most urgent necessity. Operations
of mere expediency were indeed looked askance at; they
were spoken of as 'tempting Providence', and a special
ill luck was supposed to accompany them. But they were
occasionally done for all that, and it was dreadful to see a
healthy person passing through a period of great peril or

even losing his life in consequence of some trivial operation that might have been avoided, and strange to think that it was considered justifiable to expose patients to risks which they could not appreciate, and that no one was to blame if it ended in disaster.

The most famous British surgeon at the time was the irascible Scotsman Robert Liston, Professor of Surgery at University College Hospital until his death in 1847, the year before Lister and Frank began their studies. Liston was in equal parts famous and notorious. Richard Gordon, in *Great Medical Disasters* (1983), described him as 'the fastest knife in the West End', able to sever a leg in two and a half minutes.

He was six foot two, and operated in a bottle-green coat with wellington boots. He sprung across the blood-stained boards upon his swooning, sweating, strapped-down patient like a duellist, calling, 'Time me, gentlemen, time me!' to students craning with pocket watches from the iron-railinged galleries. Everyone swore that the first flash of his knife was followed so swiftly by the rasp of saw on bone that sight and sound seemed simultaneous. To free both hands, he would clasp the bloody knife between his teeth.

There are stories about Liston which I find hard to believe could be anything but apocryphal. Reputedly he was the first surgeon to perform an operation with a mortality rate of 300 per cent. In the blur of speed while hacking into the patient he managed to slice off the fingers of his assistant and to slash the coat of a spectator. The patient and assistant died of infection; the bystander of fright. Believe it if you will. Another time, he is said to have accidentally removed a man's testicles along with his leg. H. R. Rapier, in *Man against Pain* (1947), described what happened when Liston prepared to remove a bladder stone:

> The panic stricken patient finally broke loose from the brawny assistants, ran out of the room, down the hall and locked himself in the lavatory. Liston, hot on his heels and a determined man, broke down the door and carried the screaming patient back to complete the operative procedure.

A website featuring Liston is headed '5 Insane Doctors From History Who Put House* to Shame'. But Liston was not insane. He was a practical man, and a humane one: the faster the amputation, the shorter the agony. Liston himself in 1846 performed the first operation in England using anaesthetic – a leg amputation on a man called Frederick Churchill whom he knocked out with ether. But it was to be a slow transformation, and old-style surgery took a lot longer to die than the patients who experienced it. Frank and young Lister would have learned their trade in crowded theatres, watching surgeons in 'blood-stiffened frock coats', as Richard Gordon put it, using blades which, having at best been cursorily wiped between patients, were lethal bacteriological weapons. The surgeon's toolkit might have been cobbled together from a carpenter's workshop and a butchery. Just the sight of these instruments is enough to freeze the spine. Who *wouldn't* lock himself in the lavatory? The best a patient could hope for was to faint, or to die on the spot. The crudeness of these procedures meant that the range

---

* A reference to *House*, the US television hospital drama starring Hugh Laurie as an eccentric and irascible doctor.

of possible operations was small. Surgeons could remove limbs, cut out cancers that broke through the skin (*fungating* cancers, in medical language) and take away bladder stones, but that was about it. The insides of the skull, chest and abdomen were off-limits. Such was the world of surgery as Frank first encountered it in 1848.

Bompas quoted a selection of Frank's diary entries for the first few months of 1849. He presented them without comment, observing only that they 'show both the society he enjoyed, and also how his love for natural history maintained its hold upon him'. This was fine as far as it went, but without a context it risked making Frank look more than ever like a mad scientist in the making. For example:

> *January 19.* Gave Jacko [the monkey] chloroform: very successful.
> *May 25.* – Large luncheon party. Gave the eagle, the snake, and the [gold]fish chloroform: all succeeded.

These make sense only when you remember that anaesthesia was the day's hot topic and that chloroform was being mooted as an alternative to ether. It had been a little over a year since James Simpson, Professor of Obstetrics at Edinburgh, had used it on a patient for the first time, and not much was known about its side effects. In fact a patient died under its influence in 1848, and others would have their livers wrecked. Chloroform eventually would overtake ether because it was easier to administer, but surgeons in 1849 were still feeling their way. Several pages further on, Bompas tried to explain: 'The Dean gave several luncheon parties, at which, with Frank's assistance, the effects of the new anaesthetic were tried on various animals.' But this still left the Bucklands looking more like dilettantes than serious men of science. Even so, there was no risk of public disapproval. Experiments on animals did not arouse passions in the way they do now, and Frank would have been convinced of man's God-given 'dominion over the fish of the sea, and over the fowl of the air, and over every living thing that moveth upon the earth' (Genesis 1:28).

St George's in the meantime was feeling the impact of its new arrival. As always, Frank's kindness and sense of fun guaranteed his popularity, though his kindness was not always of the sort which these days would please the General Medical Council. When a mother fretted about the amputation of her young son's thumb, Frank told her not to worry: very likely it would grow back again. When she complained that no new growth had appeared, he could only exhort her to greater patience. To another young woman he could offer no comfort at all. Her great mistake had been to trust a young man who persuaded her to have his initials tattooed on her arm. The young man turned out to be a mendacious two-timer and she now wanted to rid herself of his loathsome branding. Alas! The tattoo, complete with lovers' knot, was so huge that nothing short of amputation could have removed it. Frank's advice was to find another sweetheart with the same initials.

He noticed after a while that accident cases coming into the hospital could be classified according to the time of day. Suicides came in the early morning, followed by scaffold accidents, then street accidents. Other observations were more specific. One of his patients was an old woman who told him her cough could be relieved only by 'some sweet luscious mixture' which had been prescribed for a friend of hers. Frank gave her a bottle of the medicine but was puzzled when she returned again and again for more, while her cough did not improve. He grew suspicious and had the patient watched. She was discovered standing outside the Chelsea Hospital selling the mixture in halfpenny tarts.

Frank's diary for 2 March 1849 provides the first recorded evidence of his love of freaks. 'Saw the man who could sing both treble and bass at the same time: very curious but very sweet.' On 9 March he ate lumpfish for dinner. He found it 'very good, something like turtle', but changed his mind overnight: 'Rather seedy from the lump fish'. Away from the hospital he continued his ascent into the social stratosphere. On 25 March Sir John Pilkington entertained him to dinner with Sir Robert Peel, Lord Lincoln and the soon-to-be prime minister Lord Aberdeen. On 8 May Richard Owen (the biologist who would become the founding superintendent of the Natural History Museum) and the Archbishop of Dublin came for breakfast, and Sir Edwin Landseer arrived shortly afterwards 'to see the monkey and the eagle'. Presumably this was a different eagle from the miscreant sent to Regent's Park, though it was clearly no stranger to misadventure. A month earlier Frank had set its broken leg. At the same time he was sealing his friendship with the future superintendent of the Zoological Gardens, Abraham Dee Bartlett, who, on 30 May, 'sent me a kangaroo and a gazelle to prepare and find cause of death'.

Europe in 1849 was scoured by a cholera pandemic. In London that year it cost 14,137 lives, while Liverpool lost 5,308 and Hull 1,834. But nowhere could compare with Paris, which in 1848–9 buried more than 19,000. The city was riot-torn, overcrowded and filthy, its streets and alleys running with liquid waste. One witness described rainwater collecting in 'deadly pools infected with the organic matter of fermented excreta', a highly efficient mass-production facility for rats, fleas and *Vibrio cholerae* bacteria. In August of 1849 this was Frank's destination of choice. But it was not charity that drew him to the Hôpital de la Charité. Thanks to cholera, the attraction was a generous supply of cadavers for him to dissect. In London a freshly deceased workhouse inmate could fetch £10; Parisian bodies came much cheaper. He wrote home: 'I have been this morning around the wards of la Charité, and have begun a course of operations on the dead body; yet I have hardly been here twenty-four hours. Only think, we have a fresh subject every day, and may perform any operation we like.' Another letter quoted by Burgess struck a grimmer note:

There is a large horrid looking cart which goes round to the Hospitals in the morning and brings in the dead people. I have seen as many as 10 at one time. They are sewn up in coarse cloth. The man pulls them out and the professor chooses one to dissect for which he pays 4 francs. The teeth of these poor creatures are the perquisite of the porter and he pulls them all out to sell to the dentists. Last week I performed all the amputations and am beginning to feel confident and get my hand into the work. When I have done with the Professor I go over those operations I have done before and generally all that remains of the body is the trunk . . . The subjects are principally old men and women but we have some fine men and pretty women. The thinner they are the better.

He was also thinking about animals. 'One day I must go to Alport to see the hospital for . . . horses, dogs &c., and another day to see the horse slaughter-house.' He bought a little owl as a souvenir, and some gifts for his friends ('a beautiful skull and some books for J. Ogle').

One of a tribe: Frank's monkey Jenny

By late September he was back at St George's. We cannot know how much, if at all, his patients benefited from his work on cadavers. Despite anaesthesia, their fate might as well have been settled by a coin-toss as by the surgeon's skill. Indeed, anaesthesia might have caused even more deaths than it saved. It encouraged surgeons to perform more operations than they might have done in the old screaming-agony days, thus condemning their subjects to lingering deaths from infection.

Life at the Deanery went on much as before. On 8 October Frank set the owl's leg, and on the 11th he took delivery of a monkey, which he called Jenny. This may or may not be the ape of the same name in the previous chapter – Bompas and Frank both muddled their chronologies. Three days later: 'Monkey got loose; trouble to catch her again.' Next day: 'Put the monkey on her pole; good place for her.' Monkeys of various kinds feature throughout Frank's life – in part because of his simple fascination with them, and more problematically because of their (to him) notorious role in the theories of Charles Darwin. The intellectual and social force-fields meanwhile continued to gather around him. The guests for breakfast on 17 October were Professor Owen, the Assistant Secretary to the Treasury, Sir Charles Trevelyan, and the Prince of Canino, Napoleon's nephew, whom Frank judged to be 'very like his uncle, sharp and intelligent. He put new names to my owls.'

Frank was still only twenty-two. On 24 October he received a visit from a man called Pollock, who wanted to see the monkey. This was probably Sir Frederick Pollock, Lord Chief Baron of the Exchequer (a senior law officer), with whom Frank had dinner a few weeks later. The monkey continued to be troublesome. On 26 October it caused consternation by escaping over the roofs of Dean Street. On 19 November the social ceiling was raised yet again when 'Prince Albert came to the Abbey . . . and he talked with me about the German students'. This meeting would have its consequences. On 8 December. 'Went with Pollock to the Zoological Gardens: saw the new snake house; saw the boa eat a rabbit; heard the rattlesnake's rattle going, saw some of Cleopatra's asps, and some very young slow-worms. Owen to give a lecture next Tuesday upon the rhinoceros that has died.'

In 1851, the year of the Great Exhibition, Frank and William put on a show in Westminster Abbey, where they hung a pendulum from the roof of the nave to demonstrate the rotation of the earth. It was in this same year that 24-year-old Frank first tried his hand at authorship. In the scrapbook I find a letter indecipherably signed by the Lecturer on Physiology and Anatomy at St George's Hospital: 'Dear Buckland, I have looked over your notes of the Arterial and Nervous supply of the Muscles of the Upper Extremity and think they are likely to be prized by all Medical Students as the table is at once original and much wanted.' I suspect it was this that encouraged Frank to write a paper on the muscles of the arm, which he hawked unsuccessfully around London publishers in June, July and August. Two other things happened in 1851, both of which Bompas overlooked. The first might be passed off as an oversight, too obvious to be worth mentioning. Newspaper clippings pasted into the scrapbook record that in May of that year Frank passed the entrance examination for the Royal College of Surgeons, his passport to the career he still professed to want. The second omission is more puzzling.

Frank became a father.

On page 116 of his scrapbook, Frank himself appended a bald little note: 'Poor little Physie born Aug 23 1851', but that is all. There is no mention of the baby's mother, or of the child's real name, Francis John. Bompas on the whole was keener to describe Frank's virtues than he was to expose his warts, and in the hierarchy of nineteenth-century warts there was not much that topped bastardy. It was not the kind of thing that the scions of higher churchmen were expected to go in for. If they did, then they were at least expected to show discretion, the 'wretched girl' being paid off and dispatched to suffer her disgrace alone. This was not just a matter of social stigma (though it was that, too) or of Church-inspired bigotry (ditto): the prejudice was underpinned by law. The Poor Laws in 1834 had been reformed supposedly to encourage working-class thrift and to discourage female 'immorality' by stigmatising those who brought shame on their families. It was held that paying poor relief to mothers and their bastards was an encouragement to licentiousness. The supposed moral enforcer was the Bastardy Clause, which effectively absolved fathers of all responsibility and condemned mothers and their illegitimate children to the worst that Victorian poverty could throw at them. For those who could not support themselves, the workhouse loomed. But even here they faced a kind of moral apartheid. The Poor Law Commissioner in 1842 wanted the fallen types to be kept apart from other workhouse women, for the good of the majority. As employers generally took a similar tone, the slippery slope tipped many such women into prostitution. Reform of the Bastardy Clause in 1844 did allow them to claim recourse against their seducers but, as witnesses were required to confirm that the alleged fathers had performed the necessary deed, it was of little comfort to most young women. Killing your baby, or yourself, was an easier way out.

Let us now give Frank the credit he deserves. He did not abandon the girl. 'Poor little Physie' was his son, and he would take responsibility for him. Eventually he would marry the boy's mother. Burgess guessed that Frank first met Hannah Papps, a coachman's daughter, some time in 1850. Francis John Buckland/Papps was born at an address in Pulford Street, Pimlico. It no longer exists (the street was bulldozed in 1935 to make way for the Tachbrook Estate, now owned by the Peabody Trust), but in 1851 it ran down to the Thames between a gas-works and a builder's yard. It was within walking distance of Westminster, but the address was clearly not salubrious and Frank continued to live at the Deanery. His reason for not marrying Hannah at this time (and again I have Burgess to thank for this insight) might have been that it would have cost him the £80-a-year stipend he still received from Christ Church as a *Bachelor* of the Arts. As a student he simply could not afford it. Marriage would not follow until the August of 1863, when his financial outlook was brighter. Bompas then did acknowledge Hannah's existence, though he gave no detail and managed to get her name wrong – *Henrietta Papes* instead of Hannah Papps (her real name confirmed by the public record) – a mistake which John Upton later perpetuated. Mrs Gordon also overlooked the arrival of her nephew. It begs the question: did the family refuse to acknowledge the child and his mother, or did Frank simply prefer not to tell them? He would have had good reason.

Like corporal punishment, the stigmatisation of illegitimacy was a Victorian value that would take a very long time to fade. It persisted even in the supposedly swinging 1960s, a sword of moral terror hanging by a gossamer thread over the head of any girl who surrendered to passion or persuasion. Miserable ill-matched couples were still forced into shotgun marriages, and paternity cases, excruciating to listen to, kept local magistrates busy. Childbirth outside marriage is now so common that, if trends continue, it will become the norm. The proportion of children born to unmarried mothers in Britain in 2014 was 47.5 per cent. In the earliest recorded year, 1938, it was 4.2 per cent. No one then would have believed the current prediction that by 2016 the majority of births will be 'illegitimate'. The USA is not far behind: in 2013, 41 per cent of American first-time mothers were unmarried.

Even so, the fulminators have yet to fall silent. In Britain, politicians of the authoritarian right demand tax breaks to make marriage look a better bet financially, and a right-leaning press is periodically outraged by welfare benefits to single mothers. Thus, even now, the bewhiskered bigots of the nineteenth century might hear a satisfying echo of their distant voices. From a twenty-first-century perspective, I would argue that Frank's loyalty to mother and child put his moral account comfortably in the black. His economic account, too, started to look a bit more encouraging in 1852, when St George's engaged him as a house surgeon. At around the same time, thanks to the royal eye doctor, he made his debut in print.

The eye doctor was an old friend, ten years older than himself, William White Cooper. Bompas described him as 'the queen's oculist', but in fact it was not until 1859 that he became *Surgeon-Oculist in Ordinary* to Her Majesty. In 1852 he was most likely ophthalmic surgeon at St Mary's Hospital, Paddington. It was in the spring of that year that he visited the Deanery. 'Frank', he recalled, 'asked me to go downstairs and see his rats.' White Cooper found himself in 'a sort of cloister', where a dozen or so rats were kept in cages. Frank lifted these out one by one and 'described in a most interesting way the habits and peculiarities of each one'.

When I had seen all that was to be seen, I said, 'Frank, just you put down on paper all that you have told me about those rats, add what you please, let me have the manuscript, and I will see whether something cannot be made out of it.' Frank demurred, saying that he did not think he could write anything worth reading.

After some encouragement, he promised to comply . . . and in due time the manuscript arrived; having touched it up a little, I took it to Mr. Richard Bentley, with whom I was well acquainted, and said, 'Mr. Bentley, I am going to introduce a new contributor to your *Miscellany*; one who will strike out quite an original line.' Mr. Bentley was not greatly impressed by what I said, but accepted the MS, which appeared as an article in the *Miscellany* of the following August; and thus commenced the interesting series, subsequently collected and published as 'Curiosities of Natural History'.

White Cooper was perhaps too keen to take credit for the career of a man more famous than himself, but we should give him his due. All writers owe something to their early mentors. Frank would say that the fee – or 'honorarium' – which he received from Bentley was 'the most delightful surprise' of his life. The success of 'Rats' would lead to numerous further appearances in *Bentley's Miscellany* and in Charles Dickens's weekly magazine *Household Words*. We will come to these later, but first we need to revisit the family home.

I ought to say rather family *homes*, for the Bucklands spent time at Islip as well as at Westminster. Both communities gained much from William and Mary's philanthropy. Thanks to William, Westminster got two new churches. Thanks to Mary it got first a coffee club and then an industrial school. The coffee club was in the parish of one of the two new churches, St Matthew's, whose incumbent was a Reverend Malone. It started well but ended badly. The Reverend Malone told Mrs Gordon:

> Mrs Buckland . . . got lecturers, and among them Frank Buckland, to give weekly lectures, and a good library was formed. It answered only too well for two years, but then the police informed me it was a meeting-place for thieves, and that they formed there schemes of burglary.

Mary did not despair. She reopened the coffee club as an industrial school where street boys were, as Mrs Gordon phrased it, 'taught to make paper bags and to print; and as they were fitted for employment, they were drafted off, and many of them became useful workmen'. Mary also gave money for lending to 'the deserving poor'. At Islip, too, she became involved in the local school. Using home-made paper globes she taught basic geography and pointed to the countries where familiar commodities such as sugar and tea came from. 'Many amusing letters did she receive,' wrote Mrs Gordon, 'protesting against such unnecessary teaching, which was only supposed to put foolish notions into children's heads.' It did have unexpected consequences. The interest Mary aroused in the wider world inspired a number of families to emigrate to Australia, where, according to Mrs Gordon, they prospered as 'landed proprietors'. The dean himself provided clothes, booked their passage and 'commended them to the care of the captain'. More imaginatively, he also sent currant and gooseberry cuttings, which he packed into tin boxes filled with honey and soldered shut to keep airtight. I cannot say whether or not this was the precedent that would encourage Frank many years later to make his own contribution to Australian biology, but it must have fed his imagination.

Mrs Gordon's account of the villagers' journey to the docks is a masterpiece of gentle humour:

All the details of their journey were carefully planned and personally supervised by the Dean. Vans met the emigrants at Paddington, and they were driven to the Deanery and hospitably entertained before going down to the docks for embarkation. 'Be um aloive?' was the general exclamation as Buckland's country friends passed the Horse Guards sentries and saw London for the first time.

William's other gifts to Islip were a cottage for the 'village lads' to use as a recreation room, and the voluntary services of his son. Frank at this time was still only a medical student, but this did not stop the local doctor from filling the Rectory kitchen with 'lame, halt and blind' for the young man to treat at weekends. The length of time William now spent at Islip was not good news. In 1849 he had fallen ill with a disease or a condition from which he would not recover, and which meant he could no longer manage his duties at Westminster (these were performed on his behalf by the sub-dean, Lord John Thynne). Physically and mentally, this once brilliant and formidably energetic man gradually wasted away. His baffled doctors retired him to Islip, where he seemed for a while to enjoy his garden and the allotments he had provided for the villagers. But it was hopeless. Mrs Gordon was then only thirteen, but old enough to remember her father's 'terrible weakness, torpor, and loss of flesh'. The geologist Sir Roderick Murchison was a frequent visitor but could do nothing to rekindle William's interest in science. As Frank himself recorded, the entertainment weekly *The Leisure Hour* 'was the only publication my dear father would read during his illness, and the volumes were always on the table; he would look at nothing else, save the Bible'. In this condition it would take the stricken dean six years to die.

Though he was burdened with sadness and responsibility, Frank did not allow himself to be brought low. In 1853 he addressed his diary with all his usual enthusiasm:

February 1. – Article on Cobra in 'Bentley'.
May 26. – Took my Monkey article to Bentley.
August 30. – Took my article 'Old Bones' to Dickens.
September 12. – Wrote out article on 'Parasites' from 11 to 3.
October 10. – Made cast of girl's head, and of Ben Jonson's head; also made mould of rat's body.

October 11. – Up to the Zoological Gardens, and gave the ant-eater some ants, which he would not touch.

October 12. – Up to the Gardens with six rats; gave them, two to eagle, one to mongoose, two to rattlesnakes, and one to cobra; got poisoned afterwards in skinning the rat.

Frank was now twenty-six and lucky to be alive. The adventure of 12 October was very nearly his last. He had wanted to see how the cobra's poison would affect the rat, and made careful observations of all that happened. At first it was nothing much. When the rat was tumbled into its cage, the snake flicked out its tongue and hissed while the rat shrank into a corner and began to wash itself. But it was the rat that would make the first move, kicking off a fight that Frank recorded blow by blow like a commentator at a boxing match:

Rat provokes snake by running across its body. Cobra strikes but misses, bangs its nose against the cage and spreads out its crest. Rat darts across the cage. Snake strikes again but only half connects. Rat flies at cobra and bites its neck. Snake tries to twist and bite. Rat clings on 'like the old man in *Sinbad the Sailor*'. Snake changes tack. Lowers its head and throws rat on floor. Rat tries to run, but too slow. Cobra sinks fangs into its back.

The denouement:

This poor beast now seemed to know that the fight was over, and that he was conquered. He retired to a corner of the cage, and began panting violently, endeavouring at the same time to steady his failing strength with his feet. His eyes were widely dilated, and his mouth open as if gasping for breath. The cobra stood erect over him, hissing and putting out his tongue, as if conscious of victory. In about three minutes the rat fell quietly on his side and expired.

The cobra then moved off and took no further notice of his defunct enemy.

It was not the *fact* of the rat's death that interested Frank: it was the *means*. How had the poison worked on its body? Clearly, dissection was called for. The rat was hooked out of the cage and Frank straight away skinned it. In its flesh he could clearly see the two tiny punctures where the fangs had gone in. The effects of the poison were obvious. 'The parts between the skin and the flesh, and the flesh itself, appeared as though affected by mortification, even though the wound had not been inflicted above a quarter of an hour.' To see whether the skin itself was affected, Frank scraped it with his fingernail. Finding nothing of further interest, he threw the rat away, pocketed the knife and the skin, and set off home. He did not get far.

I had not walked a hundred yards before, all of a sudden, I felt just as if somebody had come behind me and struck me a severe blow on the head and neck, and at the same time I experienced a most acute pain and sense of oppression at the chest, as though a hot iron had been run in and a hundredweight put on the top of it. I knew instantly, from what I had read, that I was poisoned.

By good luck he was with a friend, 'a most intelligent gentleman', to whom he gave urgent instruction. If Frank fell down he was to be given brandy and *eau de luce* (a mixture of hartshorn, spirits of wine and oil of amber). At all costs the friend was to keep him moving and on no account to let him lie down. 'I then forgot everything for several minutes, and my friend tells me I rolled about as if very faint and weak . . . He tells me my face was of a greenish yellow colour.' They staggered along for a few

more minutes before Frank spotted a chemist's shop, rushed in
and called for *eau de luce*.

> Of course they had none, but my eye caught the words
> 'spiritus ammoniae', or hartshorn, on a bottle. I reached it
> down myself, and pouring a large quantity into a tumbler
> with a little water, both of which I found on a soda-water
> stand in the shop, drank it off, though it burnt my mouth
> and lips very much. Instantly I felt relief from the pain at
> the chest and head . . . After a second draft at the harts-
> horn bottle, I proceeded on my way, feeling very stupid and
> confused.

Four large brandies at his friend's house restored him sufficiently to continue home. Only then did he notice an acute pain
which seemed to begin under his left thumbnail and run all the
way up his arm. He then realised what had happened.

> About an hour before I examined the dead rat, I had been
> cleaning the nail with a penknife, and had slightly separ-
> ated the nail from the skin beneath. Into this little crack the
> poison had got when I was scraping the rat's skin to exam-
> ine the wound. How virulent, therefore, must the poison of
> the cobra be! It had already circulated in the body of the
> rat, from which I had imbibed it second-hand.

Three things strike me about this story. The first is its un-
doubted truth – William White Cooper confirmed it to Bompas.
The second is that, as White Cooper commented, Frank showed
astonishing bravery – 'pluck' was White Cooper's word for it –
and presence of mind. The third is that it exemplified par excellence the twin pillars of Frank's modus operandi – observation

and practicality. In this case they had saved his own life, but they would serve his patients just as well. At St George's he treated a boy who somehow had managed to get himself bitten by a viper in a pub. Frank made him suck the wound as hard as he could, and kept him at it for nearly two hours. He dissected the snake and found that no more than two or three drops remained in the poison glands, showing that the victim had received all that the viper could give him. Despite this the boy suffered nothing worse than a sore arm and in four days he was back at work.

Vipers were not the only things to make holes in Frank's patients. Many more were bitten by rats. Frank as usual made a careful record of all that he saw. A 'pure, punctured, clean-cut wound' was no danger to a healthy person with a good constitution, but if a drunken drayman or an overfed and pampered gentleman's servant – 'representatives of the most "unhealing" classes I know' – fell foul of a rat, then it would go hard with them. They might even die. Even so, it was better to be bitten by a rat than by a man or a horse. 'I have seen severe consequences from the former accident; and, but lately, a slave-owner in America hit his slave in the mouth, and the teeth made a severe wound, which ultimately proved fatal.'

Burgess speculated that it was during his time at St George's and the Deanery that Frank, like Sherlock Holmes with his Baker Street Irregulars, began to cultivate lowlife informers in and around the animal trade. 'Ratcatchers, bird-catchers, bugdestroyers, they were all welcomed as friends. He had learned the lesson well from his father, that the people who actually did the job had knowledge and experience that was interesting and usually worth acquiring.' He particularly valued the friendship of the zookeeper Abraham Dee Bartlett and of the animal dealer Charles Jamrach, supplier of zoos and circuses, whose shop near the docks at Wapping could provide anything from a canary to a rhinoceros. Frank copied Jamrach's habit of hanging about the docks on the lookout for sailors' exotic pets. He ruefully recounted his regret at not buying a sick-looking seal which was displayed in a tub and advertised for ten shillings. 'I afterwards heard that a Jew bought him for five shillings, cured him, and sold him at considerable profit.'

As ever, Frank remained hugely popular with his fellow students and colleagues at the hospital, where his unusual eating habits made him the butt of affectionate jokes. I find a label pasted into the scrapbook, as if from a specimen jar: *A singularly large Hydatid, extracted from a woman in St George's Hospital by Mr. F. Buckland. The rest was cooked and eaten by the operator.* In the bottom corner of the label Frank has added a scribbled note: 'One of the labels that Lloyd used to stick on my things at the Deanery'. I have to look up 'Hydatid' and find what I suspected. It is a tapeworm.

The joker was Charles Lloyd, a colleague who seems to have been living with Frank at the Deanery. Lloyd's description of Frank, preserved in the scrapbook, perhaps should not be taken too seriously, though I must confess that I did at first fail to spot the joke and let him persuade me that Frank had the physique of an orang-utan. 'As nearly as might be judged at a glance, he was four feet and a half in height and rather more in breadth – what he measured around the chest is not known to mortal man.' Photographs suggest that Frank actually was a little short of average height – I would guess around five feet four – and normally proportioned, though strongly built. Lloyd continued:

His chief passion was surgery – elderly maidens called their cats indoors as he passed by and young mothers who lived in the neighbourhood of St George's Hospital gave their nurses more than ordinarily strict injunctions as to their babies. To a lover of Natural History it was a pleasant sight to see him at dinner with a chicken before him – to watch the scrupulous delicacy with which he removed the leg out of the socket, or examined, after very careful picking, the numerous troublesome little bones which constitute the pinion, and finally to hang over him as he performed a Post Prandium examination of the head – and then to see how, undeterred by foolish prejudices, he devoured the brain.

The next surprise was not long in coming. For unexplained reasons, in June 1853, precisely one year after taking up his housemanship at St George's, Frank handed in his resignation. Burgess thought he may have been deficient in the manual skills necessary to succeed as a surgeon, but – given the dexterity Frank displayed elsewhere – that does not seem very likely. He may have been revolted by the squalor in which he had to work, or by the suffering of his patients under the knife. Again there is no evidence. Perhaps his attention was so firmly fixed on other interests that 'he found it difficult to concentrate . . . sufficiently on surgery'. Now here, one feels, Burgess may have a point.

CHAPTER FIVE

King Charles's Parrot

Whatever Frank's reason for leaving St George's, the timing was odd. Maybe Bentley's 'honorarium' had made him think he could support himself by writing. If so, the hope seems to have been dashed. He decided instead that he would like to be a surgeon in a prison. It doesn't seem an obviously attractive career choice, but Frank was sufficiently keen to persuade Lord John Thynne to appeal on his behalf to the prime minister. Lord Aberdeen's reply, dated 20 July 1853 and written on surprisingly meagre sheets of paper, is preserved in Frank's scrapbook.

Dear Lord John,

The situation of poor Buckland must naturally call for the sympathy of all who knew him, on behalf of his family. I should be glad to do anything in my power for his son; but his profession removes him very much from the sphere of my interference. I imagine the office you describe must be at the disposal of Lord Palmerston, who probably for the sake of his father, and in consequence of the good report you make of the son, would be well disposed to attend to any such request as you have made to me. I should therefore recommend an application to be made to him on the subject.

I am very truly yours,
Aberdeen

Thynne took the PM's advice and wrote to Palmerston, but the Home Secretary did no more than add Frank's name to a list. Friends in high places were all very well, but Frank was going to have to advance on merit. The prisons offering no immediate prospect, he shifted his focus to what he might have thought was the next best thing, and applied for a commission in the Household Cavalry. This would take a while to work its way through the system, but in the meantime his social star continued to ascend. In February 1854, at the precociously young age of twenty-eight and with an impressive list of sponsors, he was elected to the Athenaeum Club in Pall Mall, head office of London's intellectual elite. It seems that the one person not impressed by this was his mother, whose apparent disdain drew a scolding letter from her husband's old friend, the lawyer/naturalist William Broderip. This, too, survives in the scrapbook with Frank's own handwritten note added some years later: *A very valuable letter from dear old Broderip.*

I must confess that I could not understand your indifference and determination to stand aloof with regard to Frank's admission into the Athenaeum. I thought it of great consequence to him, exerted myself accordingly, and attended at the ballot notwithstanding an inflamed eye and that it would have been better for me to be in bed . . .

Permit me in reply . . . to remind you that in order to get employment your son must be known. It is something to have his name, education and profession suspended for a week in the meeting room of such a club as the Athenaeum with an overflow of some of the best names in London to answer for him. It is something more to be elected at a season when there has been more black-balling going on than ever I remember – nay on the very night of his election when one candidate was certainly so excluded and another I believe withdrawn.

I hope and expect that Frank is made of the stuff that is *not* fit for ledger work . . . Others as well as myself will be very much out if he does not eventually occupy a high position in his professions and it is better that he should bide his time. I should not object to a surgeoncy in the Guards to begin with.

And so it happened. On 4 August 1854 Frank was gazetted assistant surgeon in the 2nd Life Guards, stationed at Knightsbridge Barracks, on the fringe of Hyde Park, a short walk from Buckingham Palace. This was what a soldier might call a cushy billet. Though the Crimean War was blazing away some 1,500 miles to the east, the Life Guards had little more to do than guard the royal palaces and look magnificent in their scarlet tunics on ceremonial occasions. Frank's ornately scrolled certificate of appointment, issued *By Her Majesty's Command*, seems to have made the scrapbook only by the skin of its teeth. At the head of this portentous document Frank has added his own wry note: *This is dated Sept 7 1854. I actually received the document April 30th 1863. Was gazetted out April 1863.*

It would be interesting to know how Mary Buckland had provoked Broderip's testy rebuke. Reading between the lines, it seems likely that she had become exasperated by Frank's

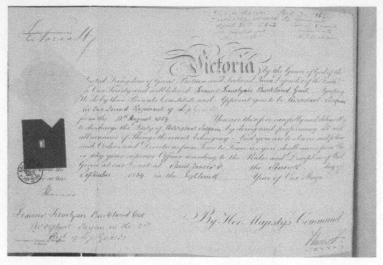

Frank's certificate of appointment as assistant surgeon to the
2nd Life Guards. It is dated 7 September 1854, but Frank's handwritten addendum shows that he did not receive it until 30 April 1863, the
month he left the army

unemployment, and perhaps had made some sour remark about
his unfitness for anything better than that icon of Dickensian
drudgery, 'ledger work'. But Frank in the army *was* taking up his
pen. He had plenty of time for it. The work of an assistant surgeon in a non-fighting regiment was not onerous, his knives and
saws rarely needed. When the guards were on parade, and during
field exercises, he had to be on hand in case someone sprang a
nosebleed or was trodden on by a horse. There were sick parades,
and soldiers' wives, children and (in the case of officers) servants
to be looked after. But really it was more like being a uniformed
general practitioner than a surgeon. It all left ample opportunity
for fishing, studying nature and turning out articles for magazines.
From the very beginning, his writing was full of funny stories.
One of his (and my) favourites dates from around this time.

Like his father, Frank habitually carried a blue bag of the kind favoured by barristers for their briefs. What travelled in it, however, was not paper but his pet monkey Jacko, whose startling habit of suddenly rolling across the floor or leaping into the air made *le bag vivant*, as Frank called it, a popular itinerant sideshow. At Southampton station one day, Jacko succeeded in forcing his head out of the bag and flashed the ticket clerk 'a malicious grin'. A furious dispute then ensued when the clerk told Frank he would have to buy a ticket for his *dog*. Nothing would persuade the clerk that the animal capering in front of him was anything so outlandish as a monkey.

A dog it was in the peculiar views of the official, and three-and-sixpence was paid. Thinking to carry the joke further (there were just a few minutes to spare), I took out from my pockets a live tortoise I happened to have with me, and showing it, said, 'What must I pay for this, as you charge for *all* animals?' The employé adjusted his specs, withdrew from the desk to consult with his superior; then returning, gave the verdict with a grave but determined manner. 'No charge for them, sir; them be insects.'

The 'malicious grin' that Jacko displayed to the railwayman was of course nothing of the sort. The alignment of lips and teeth simply reminded Frank of what such an expression would have meant in a human, and it usefully gives us a picture of what the monkey looked like. Anthropomorphism was central not just to Frank's own literary style but to most nature writing of the nineteenth century. Animals were 'he' or 'she', seldom 'it'. This takes us right back to Descartes and a tangle of contradictions and confusions which even now remains unresolved. In December 2014 the Vatican rushed out a terse denial after the Pope, while consoling a boy whose dog had died, seemed to suggest that pets might ascend into heaven. Rome hastened to place itself squarely behind Descartes. Animals had no immortal souls, which meant there could be no walkies in the hereafter. In the ancient tradition of blaming the messenger, the notion of canine immortality was attributed to ignorant journalists placing too literal an interpretation on His Holiness's soothing words.

No scientist now would write as Frank did, yet anthropomorphism remains an essential part of our culture. In one form or another – on the page, on screen, on stage, in song – humanised animals are fountains of human sentiment. It is not just in entertainment that we do this. We assign personalities as well as names to our pets, who are never 'it'. We *project*. The question we ask perpetrators of cruelty is 'How would you like it?', as if they could know what it was like to be a starving horse or a whipped dog.

Viewed rationally, anthropomorphism can look silly, a facile device for authors of fairy tales and whimsy but not for serious intellects. I think this is a mistake. What is the most powerful animal-welfare tract ever written? Not anything by the philosophical giants Descartes or Bentham but rather the work of a middle-aged Englishwoman so weak and sickly that she had to dictate the words to her mother. You could not read Anna Sewell's *Black Beauty*, first published (during Frank's lifetime) in 1877, and believe horses lacked feeling. And what satire ever bit deeper than Orwell's *Animal Farm*? We can't help it. We love animals that remind us of ourselves, and we treat them like children. An obedient dog will be rewarded with a choc drop. We humanise animals even in death, honouring them with memorials (I have one in my garden, marking the grave of a previous owner's bull terrier), and some of our grander houses even have pet cemeteries. What on earth goes on in our heads? I would argue that there is more to this than sentimentality, or rather that the sentimentality is not misplaced. The objections to anthropomorphism commonly involve what the writer Richard Kerridge has called 'reverse sentimentality', an exaggeratedly tough kind of masculinity that, *pace* Descartes, rejects the idea that animals are good for anything but exploitation. Anthropomorphism helps us to connect more directly with the creatures that share our space – 'one of our most fundamental ways of engaging with the world', as Kerridge puts it. The trick is not to dismiss it, but rather to be aware of its pitfalls. Animals are not people, and they do not see us as we see them, but we can still enjoy the warmth that a relationship with animals can bring. It is a perfect circle: good for us, and good for them. And who knows? As each new discovery is made about the ability of animals to feel and perhaps even to 'think', then maybe it is anthropomorphism, not emotionless scientific detachment, that will bring us closer to the truth.

Anthropomorphism is where childhood and adulthood share common ground. The key to Frank's character was that he never quite grew up, the boy within him never hushed. Even as official appointments and commissions of inquiry came to dominate his later life, he never lost the capacity for wonder. Everything he saw or heard would contain something to fascinate him; something that raised questions – why? how? when? – and would compel him to look for answers. Above all he loved new ideas. He kept a letter from Michael Faraday, dated 6 May 1854, asking Frank to come and see him. Frank's handwritten note explained the context: 'In answer to my asking Faraday to look at Lord Mount Edgecombe's Apparatus for ventalating [*sic*] with an Archimedes Screw.'

It was a few months earlier, in December 1853, that Frank had given his first public lecture. The venue was his mother's Working Men's Coffee House and Institute in Westminster. The subject was 'The House We Live In', a contextual pun in which the human body was compared to a dwelling. The front door was the mouth, 'with the nose as a porter, always on duty except when he has a cold'. Physiological analogies extended from the furnishings to the sewers and back up again to 'the telegraph from the mayor and council in the brain'. His debut was a huge success which led to several further performances, all enlivened by his talent to amuse. 'His drollery was irresistible, yet always informing,' said Bompas. Another persistent habit also saw its first public manifestation here. He ended with a homily:

> We are all too careless, much too careless, of our bodies. We know not, or, if we do know, we consider not, upon what slight a thread our lives depend. We are all careful of a deli- cately constructed watch, or other fragile specimen of human handiwork: how careless are we of our own much more deli- cately constructed bodies; a slight cord snaps, a small artery gives way, our real selves, the immortal part of us, our soul, in an instant quits its mortal tenement, and what remains is but clay. Let this warning remain fixed in our minds, but at the same time let us look upon our bodies, each upon his own body, and from his own body to those of inferior animals, as examples of the great Creator's handiwork, that great Creator, so omnipotent, so wise, so benevolent to all here assembled, to all mankind, to all creation.

Bompas never doubted that Frank could have made a distin- guished career in surgery. His diagnostic skills and his interest in physiology made him a perfect fit. But his passion for nature had become overwhelming. Animals were never far from his mind, or from his person. He was accompanied to the barracks by a small kleptomanic monkey called Jack (distinct from the Jacko we've already met), for which the regimental tailor made the uniform coat of a troop corporal major. The monkey's habit of picking off the insignia was punished by successive demotions to corporal, lance corporal and, finally, to private. As always, Frank's madcap humour ensured his popularity.

Much of his spare time was spent fishing for gudgeon on the Thames with 'the two brothers Reid', who were the regiment's riding master and adjutant. In his *Natural History of British Fishes*, published after his death, Frank would look back with pleasure on these days of innocent fun, covering two pages with instructions on how gudgeon should be caught. Anglers should keep the river bottom below the swim well raked (the fish were attracted by stirred-up mud), and should bait with worm. Most importantly, they should always take a frying pan, as 'gudgeons taken out of the water and immediately fried are delicious'. In conclusion, he recalled that the two biggest gudgeon he ever caught were 'in a sewer which ran along the east side of the college meads at Winchester'.

His interest in animals was well known throughout the regiment. It could hardly be otherwise. One day the colonel came looking for him while he was dissecting a mare in the stable yard. *Where's the Surgeon Major?* 'Inside the charger, sir,' said the sentry. Frank treasured a skeleton presented to him by the men, which they told him was 'King Charles the First's favourite parrot found in Windsor Castle'. He was so pleased with this that he had a drawing made of it and later displayed it in his museum. It was the perfect joke.

Its bird-like attitude, I confess, deceived me for a moment, but I quickly discovered that a trap had been laid by some of the troopers to catch 'the doctor' . . . This curious object is simply the skeleton of a rabbit, put up in a bird-like attitude; the rabbit has been cut into two, and the flesh taken off the bones, which are coloured brown to give an appearance of antiquity; the neck bones and part of the back-bone have been left attached to the head; the hind legs have also been left attached to the hip bones, or pelvis, and the two halves then fastened ingeniously together in the outline of a bird's skeleton; the hind legs have been neatly tucked up exactly like the legs of a bird sitting, and the bones of the rabbit's feet have been moistened and then turned round a perch to give the idea of a bird's claws. The whole thing was then set upon a perch to carry out the idea of a bird.

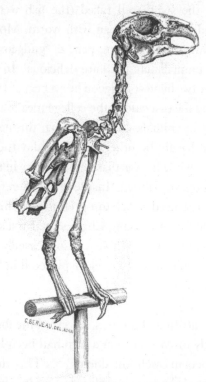

The perfect joke – 'King Charles the First's Favourite Parrot' was a spoof made by guardsmen from the bones of a rabbit

All the time Frank was writing at a prodigious rate, and with a gift for translating his whole personality into words. He was a natural. Burgess aptly described his style as 'fresh, vigorous and direct without being inaccurate or condescending'. He had the scientist's respect for facts, a novelist's eye for detail, a humorist's love of the absurd. It was the perfect package, and it would make him a star. By 1856, as Burgess observed, he was 'well on his way to becoming a well-known journalist, but not a famous scientist or surgeon'. As an example of his style, Burgess liked to quote Frank's description of a roosting bat, which he had compared to 'half a pound of brown sugar done up in brown paper and hung up by the small end'. But for the pure, distilled essence of Frank, it would be difficult to beat this paragraph from an early piece called 'Hunt in a Horse Pond'.

To the inhabitants of the pond, the pond is the world; to the inhabitants of the world, the world, as compared to space, is but a pond; and when the adventurous lizard has made a voyage of discovery round his pond, he has as much right, comparatively speaking, to boast of his performance to fellow lizards, as Captain Cook had, when he sailed round the world.

His range of subjects was both extraordinary and eclectic: sea lice, crab claws, the lobster's sense of smell, the water intake (two or three gallons in twelve hours) and loss through perspiration (six hundred gallons a year) of a Swansea furnace-man, frog showers, the bodily secretions of toads, the poisonous effects of toad poison on a drunk, cannibalism in rats, the death of rats in atmospheric railway pipes, the proper way to skin a boa constrictor, sticklebacks in the Isle of Dogs, the last

Thames salmon, the mischievous behaviour of Jacko the monkey (whose death from bronchitis he sadly had to record), foxes at Leadenhall market . . . On and on, page after page, subject after subject, a handwritten outpouring which few modern writers could equal even with all their technological advantages and immediate access to every fact known to man. Frank had to find it all out for himself, while never forgetting his duties as an army surgeon. Candles were burnt at both ends, and we begin to see how, towards the end of his life, Frank would drive himself to exhaustion.

Nothing passed unnoticed. As Bompas put it: 'Materials of every variety, his mind, singularly alert, was always gathering, and these were stored in numerous note-books.' Frank also collected newspaper cuttings, 'hoping they will turn out useful in the article line'. This meticulous self-curating was a symptom, not the substance of his genius. His retentive memory was a bottomless tank into which his restless eye poured an unstaunchable flow of facts and observations. It meant, said Bompas, that he could 'discuss and illustrate almost every subject with a readiness and resource always remarkable'.

In 1855 Frank had his first fleeting contact with the most important of his future patrons, Queen Victoria, to whom he was presented at a regimental levee. In June of the same year the Dean of Christ Church died. This was the same Thomas Gaisford who ten years earlier had given Frank the 'you or that animal' ultimatum that had sent Tiglath-pileser into exile. After the dean's rather grand funeral, Frank wandered back into Christ Church Cathedral, where workmen were filling in the grave. The verger told him that the ground there had not been disturbed for two hundred years, which immediately set Frank poking around in the soil. From the 'brickbats and rubbish' he quickly sifted out some fragments of human bone, which he tossed aside as of no interest. What did interest him was a small round bone, about the size of a shirt-stud and with a needle-sharp spike sticking out of it. This had obviously been buried for a very long time and was clearly non-human. Having examined it from every angle, Frank identified it as a spine from a thornback ray and meditated happily on 'the monkish inhabitants of Christ Church in olden time; their fast days and fish suppers, and the kitchen midden or refuse heap of the monkish cook before the days of Henry VIII, on which this fish-bone had doubtless been thrown three centuries ago'.

It was around this time that Frank began experimenting with plaster casts. In October 1853 he had recorded making a cast of Ben Jonson's head. As Jonson had been dead for more than two hundred years, we must assume that the cast was of the portrait medallion in Poets' Corner. In June 1855 Frank recorded: 'Made casts of rats, &c., in lead'. In May 1856 he 'modelled the mare's head in clay from drain' (this was the same animal inside which the colonel had found him at the barracks).

But 1856 was a black year for Frank. On 9 February, unremarked by Bompas and Mrs Gordon, Frank's son 'poor little Physie' died. Burgess tells us that the cause of death was 'probably tubercular meningitis' and that Hannah had been living with the boy at Clewer Village, near Windsor Barracks, where Frank was now stationed. A letter of commiseration pasted into the scrapbook (I could not decipher the signature) has the customary addition in Frank's handwriting: 'Poor little Physie died Feb 9 1856 aged 4½ Born August 23 1851'.

Another of Frank's notes is the only thing remaining on pages 106 and 107. Whatever he pasted here must have been lost or removed before the library received the scrapbook. The tone is likely to have been sombre. The note reads: 'Dr Buckland was made Dean of Westminster 1845. Taken ill 1849. Died

Aug 14 1856 – Dean 11 years.' For six of those eleven years William had been incapable of work, but his illness was still a mystery. 'The best medical opinions were consulted in vain,' wrote Mrs Gordon. 'The cause of the illness baffled the highest skill, but to the last it was hoped that the malady might disappear as mysteriously as it had come.' The hope was in vain. Tragically, before the illness robbed this brilliant man of his life, it first took away his reason. 'The best medical opinions' had no idea why. For all his overwhelming grief, Frank could not suppress his desire to know what had killed his father. He ordered a post-mortem.

This revealed, as Frank put it, that 'the brain itself was healthy in every respect', but the base of the skull and the two upper neck vertebrae were badly decayed. 'The irritation, therefore, communicated by this diseased state of the bones above was quite sufficient cause to give rise to all symptoms; this irritation being considerably augmented by continuous and severe "exercise of the brain in thought".' The blame for all this, in Frank's opinion, could be pinned on a German coachman who had overturned a diligence when his parents were on their way to a conference in Berlin. William had been stunned, and Mary suffered a bad cut on her forehead. 'Professor Ehrenberg, who fortunately was with them, attended to the injured, which proved the ultimate cause of death in both. Dr Buckland's vertebrae were injured, and a bony tumour was discovered at the back of the cut on Mrs Buckland's frontal bone.'

William died in Frank's arms at the age of seventy-one. Average male life expectancy at the time was 41.36,* similar to what it is today in parts of Africa. In Sierra Leone it is forty-seven years

---

* The median age at death was 46.65 and the modal age (the age at which death was most common) was 71.84.

for men and fifty for women. In Britain it is eighty-one and
eighty-two and a half years respectively, which means that our
ageing bodies are having to find new things to die of. Not many
people in developed countries now die the way William Buckland
was supposed to have done.

In the Hunterian Museum at the Royal College of Surgeons,
a curator shows me a curious relic not now on public display –
object number RCSPC/49a.5, labelled *Occipital bone, Cervical
Vertebrae, Tuberculosis, Osteoarticular*. The donor was Frank
Buckland. The bones are his father's first five cervical vertebrae
and part of his occipital bone (the saucer-shaped bone at the base
of the skull), mounted on an inscribed brass plate and resembling a tiny pagoda. If William was not entirely decapitated by
the removal of these bones, then his head would have remained
attached by only the slenderest scrap of tissue. The old catalogue
notes say the bones show evidence of osteoarticular tuberculosis,
and Frank himself described 'an advanced state of caries', imply-
ing that this was the cause of death. In fact they show nothing of
the sort. There are signs of ageing and a small chip that might
have been caused post-mortem, but nothing to suggest TB. The
gloom and 'cerebral disturbance' that William suffered are
wholly consistent with what we would now call senile dementia,
and remind me of my own mother, whose death in 2012 exemplified the slow progress of Britain's medical establishment in
diagnosing and treating dementia. My response was to write a
critical article in a national newspaper. Frank went somewhat
further. He might not be the only son to have had his father's
head cut off and his bones displayed in a museum, but he must
be in a vanishingly small minority. William Buckland, minus
his upper vertebrae and the base of his skull, was buried at
the west end of Islip churchyard, where his wife would soon
join him.

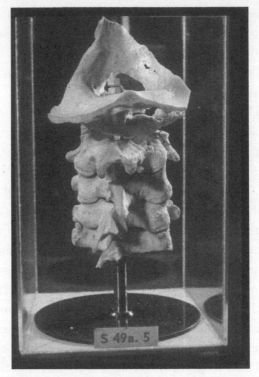

A son's tribute: William Buckland's first five cervical vertebrae and part of his occipital bone, presented by Frank to the Royal College of Surgeons after his father's death in 1856

Mary's death was recorded in Frank's diary on 30 November 1857: 'Poor dear mother died at St Leonards at 11 p.m.' For the last two years of her life she had suffered what Frank described as 'attacks of unconsciousness', which he blamed on the coaching accident. He bitterly mourned her loss. On 17 December he wrote: 'My birthday. Thirty-one years ago, at 6 a.m., I came into the world at the old house in Christchurch Quadrangle. I am now about half-way across the stage of life, and thank God I am just beginning to feel my feet. But oh! What I have lost since last birthday, the best friend a man can have in the world, his mother.' Four days later, for all his grief, it was business as usual: 'Dissected mole with view to papers in "the Field".'*

William Buckland during his lifetime had amassed a vast collection of geological and other specimens and curiosities. The most important of these – 'my geological specimens, minerals, models, maps and geological charts, drawings, sections and engravings connected with geology' – he bequeathed to Oxford University, where many of them can still be seen in the Museum of Natural History. The remaining 'Variety of miscellanies' was sold at Steven's Auction Rooms in King Street, Covent Garden, on 30 January 1857. Frank's account of this sad event was published three years later.

> [William Buckland's private collection] was transferred . . . to the dismal, condemned cell of the Auctioneer. Specimens that had been gathered by the same hand, from the same place, hundreds of miles away from home, and which had lain side by side in the same drawer many a long year, and which had been lectured on, disputed about, and admired by crowds of the most learned of *savans* . . . were now to be ruthlessly torn one from the other, destined never again to meet in their snug beds of cotton wool . . . In the auction-room I found the usual representatives of the various classes of society, deputations from the British Museum, and many learned societies of the Metropolis; from the Universities of Oxford and Cambridge; first-class dealers in minerals; and proprietors of stores in Wardour Street, and swarms of smaller 'curiosity dealers' . . . I found knowing old stagers in spectacles, who came to buy the pick of the auction . . . and green, young, ardent collectors . . . Lastly, I observed

* Frank had begun writing regularly for *The Field* in 1856.

carpenters with paper caps and aprons, who came to buy the cabinets and the book-shelves, and a good sample of dirty Jews, for without their presence a London auction cannot possibly take place.

Neither Bompas nor Burgess chose to quote this passage. The reference to Jews now makes uneasy reading, but I do not believe Frank was anti-Semitic. He was always fascinated – sometimes amused, never prejudiced – by other cultures. The nineteenth century was untouched by political correctness. Charles Dickens's *Oliver Twist*, with its stereotypical Jewish villain Fagin, had been published less than twenty years earlier by Richard Bentley, Frank's own editor at *Bentley's Miscellany*, and the prejudicial view of Jews as Christ-killing usurers was commonplace. It is right that Frank's words should now ring so discordantly.* But it is right, too, that we should see him as a man of his time and not of our own, and that we should understand the strength of his feeling. He was simply appalled by seeing his father's collection reduced to a rummage sale, picked over by people who had no appreciation of its value. I have found nothing else in his writing to suggest anything but admiration for London's Jews. Many years later, in the posthumously published *Notes and Jottings from Animal Life*, he recorded this semi-official visit to a Jewish market:

---

* The modern world should hesitate before congratulating itself too heartily. A survey of British Jews by the Campaign Against Antisemitism in early 2015 showed that more than half of them felt anti-Semitism was approaching the levels of the 1930s, and that Jews had no long-term future in Europe. Other surveys reveal that more than a quarter of British people still held a stereotypical view of Jews as money-chasers. Anti-Semitic offences in London had more than doubled in a year.

Having heard that a great number of small fry of fresh-water fish were being sent to the Jews' fish-markets in London, between Bishopsgate and Whitechapel, in the spring of 1878, I made time, during the sitting of the Committee on Mr Mundella's Fresh-Water Fishery Bill, now a most useful Act of Parliament, to inspect for myself, in order to see how far the proposed provisions of this Act might interfere with the food supply of those wonderful people the Jews. My visit was paid just before the Feast of the Passover was about to commence. This part of London is simply a Jewish colony, and the general appearance of the place and people gave me the idea of a strange foreign town. I seldom heard English spoken at all . . . but a language was used in conversation quite strange to me. I wondered if it might be Hebrew.

He went on to describe the bakers' shops selling Passover cakes, and the fried-fish shops, which he loved. Years later the Chief Rabbi himself would consult Frank on a question of Jewish dietary law, an unlikely accolade for an anti-Semite. The only Jewish innovation Frank did *not* like – indeed, pretty much the only thing he ever denounced as wholly unpalatable – was cucumber 'apparently boiled and placed in salt and water, selling at a penny or halfpenny a slice; I cannot think how anyone could eat them'. From our twenty-first-century perspective, we might register a wart against Frank's choice of words, but we can acquit him of bigotry. Two more diary entries for 1857:

January 14. – Wrote to the 'Field' about malformation in rabbit's mouth.
January 22. – Colonel Ogilvie complimented me upon my letter to the 'Field' about the rabbit's teeth, and said it was

a pity that a man who could write like that should waste his time in the Life Guards.

I don't know who 'Colonel Ogilvie' was (he was not Frank's commanding officer), but he seems to have been a man of sound literary judgement – a judgement that was about to receive enthusiastic public endorsement. In the autumn of that year Frank put the finishing touches to his first book, *Curiosities of Natural History*, containing expanded versions of his articles for *Bentley's Miscellany* and the *St James's Medley*. The preface was a clarion call for natural science:

> I heard, not long ago, of a preparer of microscopic objects, who complained that 'he had exhausted the animal kingdom'. I am well aware that many books have been written on natural history, but still am of the opinion that the animal kingdom is not yet 'exhausted', and with this belief, I have written the following pages.
>
> In natural history, as well as in other researches, it is too much the practice to copy facts and observations from printed books, the great volume of Nature herself being left unopened. It has been my great endeavour to search into this book, and to record facts which came under my own eyes . . .
>
> Without the knowledge of the structure and physiology of the lower members of the animal kingdom, it would be difficult rightly to understand many functions of the human economy; and much light has been thrown upon the art of healing by the study of the lower links of the chain of animal life . . . It has been acknowledged by many of our greatest medical men, that Natural History is the handmaid to the study of medicine and surgery.

The preface is signed *F. T. BUCKLAND, 2nd Life Guards, Knightsbridge Barracks*, and dated *Nov. 30, 1857, Athenaeum Club, Pall Mall*. Alert readers will have noted the date. Frank's revered mother Mary died on the evening of the same day.

CHAPTER SIX

Trouble with Rats

In 2002, scientists at the Imperial and Royal Holloway Colleges in London had reason to celebrate. At long last, they said, they had settled a question which had intrigued biologists ever since 'the Nobel prizewinner George Wald first drew attention to it in 1948'. They were only half right. They may have settled the question but they were adrift by nearly a hundred years in identifying its originator. Frank Buckland first raised it in 1857, the year of the Indian Rebellion. 'The cause of a lobster turning bright red when boiled', he mused, 'is a mystery I never yet heard explained. Is it mechanical or is it chemical?' A hundred and forty-five years later the answer turned out to be chemical. The science is not easy for a non-scientist to comprehend, but it involves the effect of heat on a carotenoid pigment called astaxanthin.

The point of this seemingly trivial anecdote is precisely that – its triviality. No phenomenon, however small, was beneath Frank's notice. The weirdness of the commonplace was as exciting to him as the exotica of distant continents. It is no surprise that the subject of his first published work was that most ubiquitous of mankind's antagonists, the rat. The extended version of this article in *Curiosities of Natural History* ran to ninety-two pages. Even in a bothersome rodent, he said, 'if we only look, we may find something to admire and reflect upon in the humblest works of the munificent Creator'.

He reviewed the decisive victory of the insurgent brown or Norway rat, *Rattus norvegicus*, at the expense of the 'old English' black, *Rattus rattus*, but was rightly sceptical about newspaper reports that *Rattus norvegicus* itself might be killed by super-rats from Australia. Frank himself had taken note of *norvegicus*'s penchant for cannibalism. When he left three in a cage overnight, only two were left alive in the morning. Reflecting on this, he enjoyed this story from America:

A clever Yankee being much troubled with rats . . . tried every possible plan . . . At last he got a lot of rats and shut them in a cage; they devoured one another till only a single one was left. He then turned this one loose, who, excited with the blood of his fellow rats, and having become a genuine cannibal, killed and ate all the wild rats he could find on the premises. A good Yankee story.

And it made him think. Was this not further evidence of the Creator's grand design? A wounded rat might take days to die, so it was a mercy that his comrades would 'put him out of his misery'. Nor was this the only benefit. Cannibalism, he argued, was 'a salutary check upon their increase, for a colony of rats has thus in itself the elements of self-destruction. Were all to live, there would not be sufficient food for their existence.' This puts in a nutshell the concept of what biologists now call 'carrying capacity' – the maximum population a species can sustain depending on space, food, water, and the balance between predator and prey. Frank's version may have been folksier, but it amounted to the same thing. The same thing, that is, with the important difference that it was Frank's own insight, not something learned from a book.

No experience was wasted. Recalling his misadventure with the snake-bitten rat, he developed a domino theory of rat-poisoning. A rat swallowing poison, he reasoned, would kill not just itself but also the cannibals that gnawed its carcass.[*] It was a neat solution but it lacked popular appeal. In the opinion of rat-hunting men, nothing could beat the terrier. The phe-nomenal killing power of these busy little dogs was a subject of immense pride and fierce competition between their owners. A 'Mr Shaw, of rat-catching notoriety', as Frank described him, had owned a dog called Tiny (well named: it weighed less than six pounds), which had rid the world of 2,525 rats. Had all these survived, Frank reckoned, they would in three years have produced 'one thousand six hundred and thirty-three millions one hundred and ninety thousand two hundred living rats'. His readers would have recognised 'Mr Shaw' as the London inn-keeper Jimmy Shaw, famous for the rat-baiting contests in his pub. Shaw would tip live rats into a pit, and spectators would bet on which dog could finish a set number in the shortest time. Champions had a killing rate of between 2.7 and 6 seconds per rat.

But the sport was not restricted to dogs alone. The Victorian journalist Henry Mayhew, in his classic of social observation *London Labour and the London Poor*, described a hard-up street entertainer (a fire-eater) who in desperation one day backed himself against a terrier. To ensure fairness, he had his hands tied behind his back and rushed about on his knees. The dog

---

[*] Secondary poisoning is now seen as a problem to be avoided rather than a benefit to be sought. Animals at risk from poisoned rats include dogs, cats, owls and other birds of prey, not just other rats.

went first, and set a target of seventeen minutes for twenty rats. 'Then,' said the fire-eater, 'a fresh lot were put in the pit, and I began':

> They always make an allowance for a man, so the pit was made closer, for you see a man can't turn round like a dog; I had half the space of the dog. The rats lay in a cluster, and then I picked them off where I wanted 'em and bit 'em between the shoulders. It was when they came to one or two that I had the work, for they cut about. The last one made me remember him, for he gave me a bite, of which I've got the scar now. It festered, and I was obliged to have it cut out. I took Dutch drops* for it, and poulticed it by day, and I was bad for three weeks. They made a subscription in the room of fifteen shillings for killing these rats.

To this handsome sum, having beaten the dog by four minutes, the fire-eater could add the five shillings he had bet against its owner. For a man fallen on hard times this was not a bad return for thirteen minutes' work (£1 in 1850 was equivalent to £58 now). Frank no doubt would have been aware of stories like this. His appetite for rat-lore was endless. He was fascinated by their recklessness (their flattened bodies were often found under the elephants' straw at the zoo) and especially by their terrifying omnivorousness. He recalled a gruesome case from 1851:

> The body of an unfortunate pauper, whose frame was ema-ciated to the last degree by famine and want, was brought to one of the theatres of anatomy in London . . . When the corpse was placed on the table, it was found that the whole

---

* So far as I can determine, these seem to have been salty liquorice sweets.

of the lips and parts of the ears were wanting; in place of the eyeballs were empty sockets; the parts also covering the palmar surface of the fingers were gone, only the bones and nails being left. Besides this, marks of teeth were visible on various other parts of the body.

All this had happened during a single night in a workhouse. Frank supposed that the missing parts were 'the most dainty morsels'. It was not known whether the pauper was alive or dead when the feast began, but this was not the case two years later when a baby in Bristol was bitten under the eye. 'It ultimately', wrote Frank, 'died of haemorrhage, the rat's sharp tooth having probably cut across an artery or a vein.' One may be bleakly amused by that impersonal 'it', applied to a baby. An animal in Frank's language would have been 'he'.

Frank nonetheless thought rats should be respected for their services to public health. They voraciously cleared up the 'refuse and filth' which otherwise 'would engender fever, malaria, and all kinds of horrors, to the destruction of the children of the family'. The sewers around the slaughterhouses were often choked with offal and other 'foul animal matter'. Frank challenged his readers to imagine 'what fearful maladies would arise from this putrid mass if it were allowed to stay there neglected. How is this evil prevented? Why, by the poor, persecuted rats, who live there in swarms.'

Frank himself bred so many rats that he had bagfuls of spares for the snakes at the zoo. He was especially interested in their teeth. The inner portion, he found, was of a 'soft, ivory-like composition, which may be easily worn away'. The outer part resembled hard, glassy enamel. The teeth were also beautifully clean, yellow being their natural colour and not, as in humans, the blemishing effects of tartar. Frank typically encouraged people to observe this for themselves. 'The next time the reader has a boiled rabbit for dinner, let him perform a simple experiment and convince himself of this peculiar structure common to the rat, rabbit and, in fact, to all rodent animals.' If it ever became necessary to handle a live rat, he advised, the safest method was to swing it around by the tail so that it couldn't twist and bite. To immobilise it, you should swing it with your right hand up under your left arm. 'The rat will immediately endeavour to get away, and so doing fix himself on your waistcoat; bring your arm to your side, and you have him a prisoner.'

He had sage advice, too, on how to purge rats from ships. Aside from eating them, as British expeditionaries in the Arctic had done, the usual method was to smoke them out. But, said Frank, 'a rat is not so easily killed by stinking vapours'. He preferred a novel system in which steam was piped through a leather hose. 'The rats can't stand this; they get parboiled, and the cockroaches with them; the ship is thus thoroughly freed from all its live stock.' For houses he recommended catching a rat alive, smearing it with tar or turpentine, then releasing it to stink out the runs and drive the other rats away.

All this filled Frank with delight. He had the childish habit of taking for granted that everyone else would share his excitements – which, given the power of his writing, they usually did. He made extraordinary assumptions about his readers' capabilities. 'Let the reader examine the pectoral or breast muscles of the next gull he kills, he will find them one solid mass of firm hard muscle admirably adapted to sustain and work the wings. What models of beauty and lightness are the wings of the gull!' His description of them is as beautiful as it is precise:

The bones are composed of the hardest possible kind of bone material arranged in a tubular form, combining the greatest possible strength with the greatest possible lightness. If we make a section of the wing bone of a gull, or, better still, that of an albatross, we shall find that it is a hollow cylinder like a wheat straw; but in order to give it still greater strength, we see many little pillars of bone about the thickness of a needle extending across from side to side: these buttress-like pillars are in themselves very strong, and do not break easily under the finger. Again, at the top of the bone we find two or three holes which communicate with the interior: through these, when the bird is alive, pass tubes which are connected with the lungs; so that when the bird starts for a flight, he fills his wing and other bones with air, causing them to act somewhat like a balloon on either side of him.

Thus we see not only why a bird can fly, but also why a man cannot. Sometimes the adult Frank could behave even more recklessly than the Winchester schoolboy. What boy can resist a frog? Not Frank. He had often seen boys, for a halfpenny 'dare', swallow live frogs. This was a gap in his own experience, so, purely for the sake of research, he tried it for himself. Afterwards he declared himself no better or worse for the endeavour, though 'I am told that my "croaking fits" date their origin from the moment poor frogee entered upon his fatal journey'. While stunts like this could be amusing, they also provided ammunition for critics who thought Frank should not be taken seriously. No man ever wrote about natural science with more serious intent than Frank did, but for those who confused pomposity with seriousness it was an easy mistake to make. Clive James had it right. Common sense and a sense of humour 'are the same thing, moving at different speeds. A sense of humour is just common sense, dancing. Those who lack humour are without judgment and should be trusted with nothing.'

Improbably it was a box of frogs that brought Frank as close as he would ever come to squeamishness. He loved eating French frogs, which he likened to whitebait, but he did not like the 'long price' in Parisian restaurants and so decided to buy some raw. At the market in the Faubourg Saint-Germain, a 'stately-looking dame' at a fish stall produced a box of live specimens at two a penny. What followed was a tableau of squirmy horrors. The proprietress 'dived her hand in among them, and having secured her victim by the hind legs, she severed him in twain with a sharp knife, the legs, minus skin, still struggling, were placed on a dish; and the head with the forelegs affixed, retained life and motion, and performed such motions that the operation became painful to look at'. So far as I know, this was the only recorded instance of Frank actually averting his gaze. Even so, sensitivity soon yielded to hunger. 'These legs were afterwards cooked at the restaurateur's, being served up fried in bread crumbs, as larks are in England.' This time it was not whitebait but a dish of rabbit that they reminded him of. He did sample English frogs, but found the flesh much less tender than the French. 'Should any person wish to have a dish of real French frogs, he can buy at Fortnum and Mason's, for half-a-guinea, a tin-caseful.'

He also tried boa constrictor, which he thought tasted like veal. More practically, he had a useful tip for anyone required to save a friend from the coils of a boa. *Don't tug at the snake's body. Grab the tip of its tail and you'll find it easy to unwind.* He had a keen interest in snakes, and was scornful of the misconceptions spread by writers and artists who couldn't tell a cobra from a tapeworm. Shakespeare was excoriated for supposing that snakes 'sting' with their tongues, and medical writers for vesting snakes with magical powers. He came down hard on the

seventeenth-century German Jesuit Athanasius Kircher and his
followers for promulgating the myth that snakes in La Grotta
del Serpi, near Bracciano in Italy, could cure life-threatening
diseases.

Like his father, Frank was always hot on frauds, though he
was cheerfully tolerant of the penny-a-view showmen who made
money from them (indeed he was one of their best customers).
His analysis of a 'fossilised man' at Hungerford market on the
Strand would have done credit to Sherlock Holmes. The relic,
he found, was not a fossil at all but rather the dried-up body of a
pigtailed sailor called Christopher Toledo who had died in 1670
and been 'buried in guano'. 'This poor man', reported Frank,
'. . . had met with cruel treatment; he had probably been tor-
tured and then murdered.'

> The right collar-bone had been broken: and that he had
> lived some time after the accident, or more probably the
> blow, was evident from the fact that his right arm was much
> larger than the left – the result of swelling. Had he died
> immediately, the arm could not have had time to swell. I
> thought the teeth projected in a remarkable way. I exam-
> ined them more closely, and ascertained that the upper lip
> had been cut off quite to the bone. The owner of the skel-
> eton was not aware of this until I pointed it out to him.

Another showman advertised a 'Hottentot adder', an ingenious monster in the tradition of King Charles's parrot. Its head
was constructed from eel skin, with glass eyes and spines from
a skate stuck to its forehead. The mouth and tail were from
the same fish, and the feet from a tortoise. Frank loved the fact
that the showman sincerely believed the thing was genuine. In
July 1857 he visited Cremorne Gardens, the grand pleasure

gardens which then stood by the Thames near Cheyne Walk in Chelsea, to witness a demonstration of human magnetism by a twelve-year-old girl on her brothers. 'Great humbug . . . examined the people when in magnetic state; not uniformly stiff; the iron-lifting done entirely by knack. The people go determined to believe what they see.' Like his father, Frank wanted to distinguish the real from the imagined. Strangely for one so convinced by biblical truths, this led him to think about the serpent in the Garden of Eden. A stickler for accuracy, he was annoyed by depictions of snakes undulating vertically instead of sideways. But it was not just biological accuracy that bothered him: it was logic. In Genesis 3:14, the guilty serpent learned its fate: 'And the LORD GOD said unto the serpent, Because thou hast done this, thou *art* cursed above all cattle, and above every beast of the field; upon thy belly shalt thou go, and dust shalt thou eat all the days of thy life'.

Frank reasoned that there would be no point in God condemning the serpent to crawl on its belly if that is what it was doing already. The implication was that snakes hitherto had moved in some other way, which they would now have to forfeit. On legs, for example. So all those paintings of limbless snakes in the Garden of Eden were not just wrong, they were stupid. But there was a problem. The Old Testament's authors were clearly unaware of any creatures whose time preceded their own. They made no mention of dinosaurs, nor of any other precursors of existing birds or beasts. Nor was there any evidence of pre-biblical proto-serpents. 'All the pre-Adamite snakes whose bones have hitherto been discovered', wrote Frank, 'have certainly the same conformation of vertebrae . . . as those of the present day.' In need of guidance, he consulted a 'learned divine' who, having 'looked into commentators ancient and modern', failed to deliver a verdict. Frank would have to work it out for himself.

He argued that if the pre-Adamite serpent had moved with its belly *off* the ground, then it would have needed a very different kind of backbone. Indeed, it must have been a different animal altogether. But what? His eye fell upon lizards, and in particular upon the southern African *Saurophis*, or snake-lizard. This was 'a curious creature that seems intermediate between a snake and a lizard. It has four minute limbs, which are furnished each with four toes . . .' *Saurophis* is not an animal now recognised by science, but Frank's description closely fits *Tetradactylus africanus*, the African long-tailed seps, which has tiny legs but looks and moves like a snake. He thought it 'within the limits of possibility' that this was how the serpent had looked before God's curse. The boa constrictor also gave food for thought. Like pythons, boas have what look like tiny leg bones hidden within the muscles near their tails. Having no obvious function, they are described as *vestigial* – relics of once useful limbs now shrivelled by evolution. Frank thought differently. It is important to remember that he was writing at least two years before Darwin published *On the Origin of Species*, and three years before Thomas Henry Huxley coined the term *evolution*. Though Frank himself believed the stumps were leftover pre-Adamite legs, he placed them squarely within the confines of Genesis 3. After the curse, he reasoned, the limbs would have been taken away, but the vestiges could have been left as a reminder of the snake's fallen condition. The rather delicious irony is that creationists now completely understand the evolutionary implications of these tiny bones, and furiously deny that they were ever legs at all. (Have a look at the Creation Today website if you want to see the spittle fly.)

None of this deflected Frank from his experiments. He dissected snakes' heads and wrote detailed descriptions of how the poison was made and delivered. The venom, he discovered, was a transparent, yellowish fluid of about the same consistency as saliva. It was odourless, with *a sharp taste*. This discovery was not as risky as it sounds. It was already well known that venom was safe when swallowed. The Prussian explorer Alexander von Humboldt had witnessed a man ingest all the poison from four large Italian vipers and suffer nothing worse than a foul taste. Were it otherwise, the first-aider's advice to suck venom from a snakebite would be an invitation to suicide. Frank put it nicely. By 'a wise provision of Nature', he said, 'no external poison is an internal poison, and *vice versa*'. All the same, you have to be pretty certain that your gums are in good order: one tiny nick and you're dead.

In May 1857 Frank somehow got hold of a dead puff adder. He put its venom under the miscroscope and was struck by the beauty of its hair-like crystals. But what he really wanted to know was whether it still retained its killing power after the snake had died. His unlucky laboratory assistants were a guinea pig and a sparrow, both of which he injected with the poison. The guinea pig suffered only a brief spasm of wheeziness, and the sparrow took twenty minutes to die, suggesting the four-day-old venom had lost its potency. At about the same time, the librarian of the Royal College of Surgeons, Thomas Madden Stone, gave Frank some 'wourali', now better known as curare, the plant-derived arrow poison from South America. The outcome this time was very different. The poison was thirty years old but still lethal. Frank bought a thrush in St Martin's Lane and dabbed a watered-down solution of wourali onto the exposed muscles of its breast. Alas, poor bird. After three minutes it began to pant and gape, and the nictitating membranes, the translucent third eyelids that protect the eyes, closed up. Three minutes later the bird sprang into the air and was dead before it hit the floor. Frank repeated the experiment with frogs, which died between five and twenty minutes after poisoning, and a rat. It was the rat that most interested him. As the cause of death with curare is asphyxia, then it ought to be possible to keep victims alive by artificial respiration. In this way Frank managed to keep the rat going for an hour and twenty minutes and was disappointed not to have done better. '[It] so far recovered as to be able to crawl along, but it soon afterwards died; I had great hopes I should have recovered it.' Exploits like these tended to increase Frank's reputation for eccentricity. They ought rather to have cemented his reputation for original research. He was always on the lookout for something useful to know. How, for example, did Indian snake charmers survive their close encounters with cobras? Frank soon found the answer.

> I have requested a friend in India to report to me his observations on the tame cobras. He tells me that the men make a cut under the upper lip of the cobra they wish to tame: they then turn up the lip and expose the parts which secrete the poison, or rather the duct and the reservoir for the poison just above the teeth: this they completely eradicate with a knife, and then apply the hot iron, effectually destroying the parts. A snake thus operated on would be for ever afterwards harmless, as the burnt parts could never recover themselves.

Nevertheless, he continued to regard 'miracles' with an open mind. The bandmaster of the 2nd Life Guards, C. F. H. Froehnert, had once seen a man in India bitten by a cobra and saved by a ball and a stick. The pea-sized ball had been pressed directly onto the wound, while the stick had been used to stroke the man's arm from the shoulder towards the fingers, in the opposite

direction to the poison moving through his bloodstream.
Within five minutes he had started to recover. Froehnert
was so impressed by this that he asked to be given two of the
little balls and the stick. Some time afterwards he applied
them to another victim of snakebite, and once again the man
survived. Back in England he gave his trophies to Frank,
who identified the stick as 'a little bit of common gamboges'.
This was often used as a purge but was not recognised as
an antidote to poison. Frank examined the balls with John
Thomas Quekett, conservator of the Hunterian Museum,
but all they could agree upon was that it was 'composed
of some vegetable material'. What were they to conclude?
Frank resisted any temptation to make a joke. Instead he
constructed an argument for what would now be called
'biological diversity'. Not for the first time, and far from the
last, he was way ahead of the curve.

It is an unwise and unphilosophical act to laugh at and
leave unexamined as good for nothing the native remedies
of any country, and I am quite certain that there are many
substances, particularly in the vegetable kingdom, which
are successfully used in the cure of disease by the natives
of India and other countries, long experience having taught
their value; yet these remedies remain unknown to our
Pharmacopoeia simply from want of observation on the
part of Europeans.

Frank was always attracted by oddity, and displayed a com-
passionate interest in human freaks. His diary for 10 July
1857 found him visiting 'the nondescript or hairy woman at
Regent Street Gallery'. Next day he was at the Royal College
of Surgeons, where he 'worked hard collecting material for

article on the hairy woman'. His subject was the unfortunate Julia Pastrana, known variously as the Nondescript, the Bear Woman or the Ape Woman. She had been born in Sinaloa, Mexico, in 1834, and her appearance was truly grotesque. By a perfect storm of ill luck she was only four feet six inches tall and suffered from a genetic condition, hypertrichosis lanuginosa, which covered her face and body in thick black hair, and a rare disease, gingival hyperplasia, which so thickened her gums and lips that she looked like a gorilla. Neither affliction was then medically understood, thus condemning her to wild and cruel speculation. Was she even human? Throughout America and Europe freaks were big business, often traded as chattels. Julia's early history was obscure, but by 1854 she was being managed by a man called Beach, who exhibited her at the Stuyvesant Institute in Manhattan as 'The Hybrid, or semi-human Indian from Mexico'. An advertisement in the *New York Times* identified her as the 'link between mankind and the ourang-outang'.

Charles Darwin himself took note of her, though his interest was not in her evolutionary niche but rather in the perceived relationship between abnormal hair and bad teeth. He first heard about Julia from Alfred Russel Wallace, who had himself learned of her from a London dentist called Purland. This transparent crackpot was described by William Bryant in his entertaining life of Wallace, *The Birds of Paradise*, as a 'mad tooth-puller' whose patients' shrieks were insufficient to dent his faith in the superiority of hypnotism over anaesthetics. According to Bryant, Purland was a collector of Greek coins, young boys and Egyptian antiquities. I mention this only to underscore two important facts. First, that cranks were not only commonplace but could be taken seriously even by Wallace and Darwin. Second, that Frank was not one of them.

'The Ugliest Woman in the World': the much-travelled Julia Pastrana

In 1855, ownership of Julia Pastrana passed to a man called
Theodore Lent, previously an agent of Phineas T. Barnum. He
married her and at once set off on a tour of Europe. It was during the course of this that Frank would meet her, two years later
at the Regent Street Gallery. Unlike the voyeurs who regarded
her as an animal, he saw beyond the grotesquery to the person
trapped inside. Years later he would write:

I well recollect seeing and speaking to this poor Julia
Pastrana when in life. She was about four feet six inches in
height; her eyes were deep black, and somewhat prominent,
and their lids had long, thick eyelashes; her features were
simply hideous on account of the profusion of hair growing on her forehead, and her black beard; but her figure
was exceedingly good and graceful, and her tiny foot and

well-turned ankle, *bien chaussé*, perfection itself. She had a sweet voice, great taste in music and dancing, and could speak three languages. She was very charitable, and gave largely to local institutions from her earnings. I believe that her true history was that she was simply a deformed Mexican Indian woman.

In 1859 in Moscow, Julia bore a son who inherited her hypertrichosis and died almost immediately. Julia herself died five days later of postpartum complications at the age of twenty-six, but the doubly bereaved Theodore Lent did not miss a trick. He went on touring with his family's embalmed bodies, Julia dressed in a Russian dancer's outfit and the baby in a sailor suit. Thus it happened that in February 1862, Frank would meet her for a second time. He had been invited to examine 'a great natural curiosity described as "The Embalmed Nondescript", then being exhibited at 191 Piccadilly', where it was propped up on a table. He recognised her immediately and was astonished by the skill of the embalmer.

The limbs were by no means shrunken or contracted, the arms, chest, &c. retaining their former roundness and well-formed appearance. The face was marvellous; exactly like an exceedingly good portrait in wax, but it was not formed of wax. The closest examination convinced me that it was the true skin, prepared in some wonderful way; the huge deformed lips and the squat nose remained exactly as in life: and the beard and luxuriant growth of soft black hair on and about the face were in no respect changed from their former appearance. There was no unpleasantness, or disagreeable concomitant, about the figure; and it was almost difficult to imagine that the mummy was really that of a human being, and not an artificial model.

The embalmer had done an even better job than Frank realised. Poor Julia went on being traded after her husband's death, and as recently as the 1970s was still being displayed at Norwegian fairgrounds. In 1976 thieves stole her from a warehouse and dumped her in a waste bin, whence she was removed to an anatomical collection at the University of Oslo. By then she had been the subject of a film, Marco Ferreri's *The Ape Woman*, made in 1968. A play, Shaun Prendergast's *The True History and Triumphant Death of Julia Pastrana, the Ugliest Woman in the World*, followed in 1998, when it was performed in London. Thus began Julia's slow process of redemption and reconciliation. A later production of the play at Fort Worth, Texas, in 2003, was produced by Kathleen Anderson Culebro with costumes designed by her sister Laura Anderson Barbata. When the play transferred to the Greenwich Street Theatre in New York, the sisters decided that Julia's indignity had gone on long enough. In February 2013, after a long campaign and 153 years after Julia's death, Barbata succeeded in having her body returned to the Mexican town of Sinaloa de Leyva, where she was buried in a white coffin strewn with roses. The state governor, Mario Lopez, begged his listeners to 'imagine the aggression and cruelty of mankind she had to face, and how she overcame it'. It was, he said, 'a very dignified story'. The BBC meanwhile remembered Julia with cruel simplicity as 'the world's ugliest woman'.

By 1857 Frank had begun to think seriously about fish. He was intrigued in particular by their uncanny instinct for water. 'How is this to be accounted for?' he wondered. 'I have placed a fish where he can neither see nor hear the water, even supposing that he has the power of seeing and hearing when out of his own element; yet he has always jumped in the proper direction towards it.' Despite this innate urgency, fish seemed also to have a peculiar ability to survive on dry land. To demonstrate this, Frank kept some goldfish wrapped in damp grass overnight. Half of them survived for a while, and the other half (waste not, want not) he cooked and ate, though he found them 'like carp, and I cannot say good eating'. He had also begun to worry about polluted water. No salmon had been caught in the Thames since 1833 (the last fish was presented to William IV, who paid the netsman a guinea a pound for his 'dainty dish'), and now there was a shortage of every kind of fish in the Serpentine. Frank was convinced that this had something to do with the filthy water. 'This summer I bought three good-sized bream, from a man who had caught them in this dirty pond, and I had them cooked for dinner: they tasted so strongly of the foul mud, that to call them simply "nasty" would be a stretch of politeness.'

His sense of humour continued in good shape. He laughed long and loud at a newspaper report of eels in the fountains at Charing Cross. The writer deduced that they had escaped from the fishmongers' shops in Hungerford market and had been led by instinct towards the nearest water. Frank was delighted with this glorious compendium of errors. First, the fountains were not the nearest water: that honour belonged to the Thames. Second, only an idiot could visualise an eel escaping from a shop, dodging the packed traffic in the Strand and then climbing the sides of a stone basin. Third, Frank himself knew exactly how the eels had got there. Four years earlier he had put them there himself.

By 1858, though still serving in the army, he had confirmed his literary reputation. One might add 'for better or worse'. In the nineteenth century, the worst sin a scientific writer could commit was to be popular. Such was the public's appetite for his work that 1858 saw two new editions of *Curiosities of Natural History* and a revised edition of his late father's *Bridgewater Treatise*. This latter received warm if rather quaint praise from Richard Owen at the British Museum, who thought it 'the best elementary book

Substance over style: Frank dressed for work

that a country gentleman or azure lady could take up for those sciences. It very honourably associates your name with that of your father.' The strait-laced bores of the scientific establishment persisted in dismissing Frank as a gadfly. Their objections were compounded by his rising popularity as a lecturer, and by his habit of entertaining his audiences with props and anecdotes instead of droning at them in the approved monotone. In February 1858, propitiously as it would turn out, he delivered his first lecture at the South Kensington Museum (now the V&A), one of a series given to 'working men' by a stellar cast including Richard Owen and T. H. Huxley. These were the documentary headliners of their day, the pre-televisual equivalents of the Attenboroughs and Schamas who could attract huge audiences. Frank's subject, evoking memories of his unorthodox contributions to the Oxford debating society, was 'Horn, Hair and Bristles', which he illustrated with, among other things, a hank of hair from the prizefighter 'Dutch Sam'.* He was nervous at first, but soon hit his stride. 'A heap of people there,' he wrote, 'who all said they were much pleased. Received many congratulations afterwards. Am thankful for success.' He kept a clipping from the *Windsor and Eton Express* reviewing a repeat performance of 'The House We Live In' at the Windsor Working Men's Association, which gave a graphic account of his style.

> Among the specimens a New Zealander's head with the face tattooed, a large shell from the China Seas, a rat with over-grown teeth, the vertebrae of a boa constrictor, a bone tied in a knot, a monkey's skeleton, a model of an Etruscan tomb attracted much attention. The lecture was listened to with marked attention, and was frequently applauded throughout.

In the following months he would give further performances of 'The House We Live In' at the Mechanics' Institutes of Abingdon, Newbury and Wantage, and decided he performed best when he abandoned his notes and spoke off the cuff – another black mark in the scientific book of infamy. Facts and hypotheses spilled from him like spray from a wet dog. What was the weight of a head of human hair? *Half to three-quarters of a pound.* Good-quality hair for wig-making, he reported, could fetch between thirty and sixty shillings a pound. The most sought-after, and hence the most expensive, were the blonde tresses of Germany

---

* Samuel Elias (1775–1816), known as 'The Man with the Iron Hand', reput-edly the inventor of the uppercut.

and Scandinavia. 'The darker shades are supplied by the female peasants of France, from whom the spring harvest of hair is stated to average 200,000 lbs [89.23 tons].' He noted after a visit to an ivory warehouse that rats had a marked preference for African tusks over Indian. Why was this? *There was insufficient gelatine in the Indian ivory to be worth the trouble of gnawing it.* Makers of billiard balls had the reverse preference. Frank explained: 'The African ivory contains more moisture and gelatine than the other kind; and this occasions the ivory to shrink in drying, and consequently the ball . . . to get out of the round [lose its shape].' The test of a good ball, he advised, was to roll it gently along a billiard table until it stopped. If it then made any kind of lateral movement it was useless.

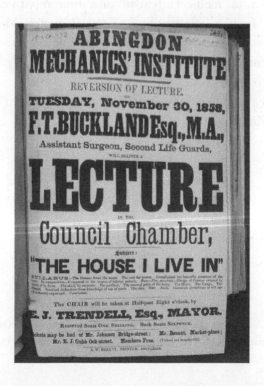

In his thinking about ivory he was both of his time and ahead of it. He was *of* his time when he wrote that 'the most nutritious jelly that we know of' was made by boiling up the dust from ivory sawn for cutlery handles. *Ahead* of it when he considered the likely impact of the ivory trade on elephants. Imports from India and the East over the previous seven years had averaged 3,341 hundredweight, or 167.05 tons. 'When we come to consider that each elephant has but two tusks, and how many must be killed to make up the above weight, it will be easily seen how it is that elephants are gradually getting rarer.'

The year ended disappointingly. In November Isambard Kingdom Brunel, then suffering from Bright's disease,* invited Frank to accompany him – 'all expenses paid and a guinea a day besides' – as his medical attendant on a three-month expedition to Egypt. Frank's commanding officers, Colonels Martyn and Vyse, gave him leave but the senior surgeon objected that his absence would 'put him to great inconvenience', so Frank had to decline the offer.

Happily, Brunel survived without him and brought him back a stuffed ibis.

---

* A kidney disease now known as chronic or acute nepthritis. Other sufferers included Tsar Alexander III of Russia, Bram Stoker, Gregor Mendel, Jean Harlow, Victor Trumper and Emily Dickinson.

CHAPTER SEVEN

Vagabond Curs of the Ocean

*I*6 *January 2015*. I am looking around for a table to park my lunch. On my tray is an enormous bap bulging with cheese, pickle, plum tomatoes and lettuce (£3.95), a raspberry and white chocolate muffin (£2.85) and a large cappuccino (£2.35). These are very friendly prices for a sit-down in central London, so it's hardly surprising the place is filling up. On the menu are sandwiches, salads, cakes, soups, puddings, hot mains of the day (beer-battered cod and chips; roasted pepper, aubergine and fennel ratatouille), hot and cold drinks. The atmosphere is as discreet as the lighting: a quiet susurrus, no raised voices, smart young staff hovering to whisk away the crumbs. A poster advertises Wednesday Jazz Night. Only by glancing up at the vaulted brick roof on its pale stone piers do I get a tangible reminder of where I am. This is the crypt under St Martin-in-the-Fields, Trafalgar Square, now rebranded the Cafe in the Crypt. The subterranean cavern also contains a gift shop, lavatories and a box office selling tickets for concerts in the church.

It was in this very place exactly 156 years ago that Frank Buckland came in search of the second great hero of his life, John Hunter. 'Hero' hardly does justice to the enormity of Frank's regard for this man. In the history of medicine, Hunter stands as an all-time great, the founder of scientific surgery, pioneer of

anatomical dissection, blood chemistry and tissue grafting. In
my own somewhat partial view, not the least of his achievements
was to have been the model for Frank's own way of thinking.
Both were benignly but determinedly ruthless. If Hunter had a
question, then there was nothing – not scientific convention, nor
even the law – that would stop him from seeking the answer. To
know a little of Hunter is to know a great deal about Frank. It
was Hunter who legitimised Frank's contempt for texts. Born in
Lanarkshire on 13 February 1728, he spent his childhood pes-
tering the grown-ups with questions about clouds and grasses,
the colour of autumn leaves, the lives of ants, birds, bees, tadpoles
and caddis worms – all things which, as he would say himself,
'nobody knew or cared anything about'. Aged twenty, he joined
his elder brother William's anatomical school in London, where
he helped with dissections. William's attempt to have John educated at Oxford foundered within weeks. 'They wanted to make
an old woman of me,' John complained, 'or that I should stuff
Latin and Greek at the university, but these schemes I cracked
like so many vermin.' Everything that followed from this – in his
own life and in Frank's – arose from this stubborn preference for
observation over page-turning.

Listed in full, Hunter's achievements would double the
length of this chapter. Early discoveries included the function
of the lymph glands and the placenta, and the fact that the tes-
tes descended in the eighth month of pregnancy. In 1762–3
he served as a military surgeon in the Anglo-Spanish war and
became expert in gunshot wounds. It does not seem too fanciful
to suppose that this played at least some small part in Frank's own
decision to enlist in the 2nd Life Guards. It was while serving in
Portugal that Hunter began the collection of lizards and other
creatures whose preserved bodies, running eventually to more
than 13,500 anatomical, biological and pathological specimens,

would fill his teaching museum in Leicester Square. In 1767 he was elected a Fellow of the Royal Society – a covetable honour which Frank himself would try but fail to attain – and in 1768 he became surgeon to St George's Hospital, where eighty years later Frank would enrol as a student. What set Hunter apart was his experiments. Unlike his contemporaries he used animal as well as human anatomy to learn about the structures and functions of the body, and it was this that made his reputation as the father of modern surgery. He set another precedent for Frank in acquiring dead animals from menageries and using them in experiments. The procedures were inspired but sometimes macabre. One of his most famous involved grafting the spur of a fighting cock onto its head. His most notorious involved boiling a giant.

The story of the giant varies according to source. Some give his name as O'Brien, others as Byrne,* but it is agreed that he was Irish and extremely tall. It is common ground, too, that he worked with a showman called Joe Vance, who advertised his height as being in excess of eight feet, and who exhibited him until the giant's death, aged twenty-two, in 1783. Charles Byrne, as I shall now call him, knew very well that Hunter would try to get hold of him when he died. Anxious to avoid this indignity, he is supposed to have paid for burial at sea. How Hunter managed to get round this is a matter of controversy. The more lurid versions have him bribing the undertaker to switch Byrne's body for the equivalent weight of rocks, with estimates of the bribe ranging from £500 to £800, both enormous sums. More sober accounts suggest that he simply bought the body for £130, though it is not clear from whom. What is certain is that Hunter

---

* O'Brien in fact was Byrne's stage name, enabling him to claim descent from the High Kings of Ireland.

did get hold of Byrne's body and carted it to his house at Earls Court,* where he boiled off the flesh in a copper boiler.

In the 1860s or 70s Frank made a pilgrimage to the house, which by then belonged to Dr Robert Gardiner Hill, a controversial surgeon who ran the place as an asylum for women. It could hardly have seemed any madder than it was in Hunter's day. Wolves, jackals and dogs had been chained up in the great man's stable yard. Lions and leopards had been caged in vaults, whence the leopards one day escaped and fought the dogs (Hunter quelled the riot but the excitement made him faint). By the time of his visit Frank was a distinguished man in his own right, but it was the inner boy who gaped at this snapshot of his hero's life. In the house he contrasted the sparseness of Hunter's old rooms with the ornamental excesses of his wife's. 'No doubt', he said, 'John Hunter had as great a horror of feminine interference in his studio as have most philosophers of the present day.' He then moved on to the covered cloister:

I have no doubt . . . that in this cloister he kept many of his smaller animals used for experiments, such as dormice, hedgehogs, bats, vipers, snakes, and snails, for his researches on torpidity; and hutches full of rabbits, whose unfortunate fate would be to have their ears frozen, to prove points connected with blood circulation.

It would also be a good place to hang up skeletons, or dry preparations, or to macerate bones . . . The entrance into these cloisters leads through a subterranean passage,

* In 1873 Hunter moved to a large house in Leicester Square, where he took in pupils and established his teaching museum. It was said that resurrection men (i.e. body-snatchers) came at night with bodies for the dissection rooms.

very dark, and like an enlarged fox's earth. This passage, again, I warrant, was one of Master John's contrivances, for into his burrow he could wheel a tidy-sized cart or truck, and drag into his den anything from a giant's body up to a good-sized whale . . . John could have easily whipped anything into the stable-yard down his fox's earth, and into the area, without Mrs. Hunter knowing anything about it; and I'll be bound to say she used occasionally to 'lead him a life', and kick up a row if any preparation with an extra effluvium about it was left on the dissecting-table, when the great surgeon was obliged to go out on his professional duties.

In a small room at the end of this burrow he felt a rush of excitement when he spotted some bones, though they disappointingly turned out to be kitchen waste. And then he saw it – 'a largish-sized copper boiler standing out of the wall . . . Ah! If this old boiler could only tell what it had boiled! *One* giant, we know, was boiled up in it.' Byrne's enormous skeleton was exhibited first at Hunter's museum in Leicester Square and then at the Royal College of Surgeons, where it still remains. But it was not the giant's bones that were on Frank's mind in 1859: it was Hunter's own.

The great surgeon died on 16 October 1793, aged sixty-five. He had been a disputatious man who died as he had lived, suffering a fit during an argument with the board of governors at St George's. A few days later he was interred a few feet from where I now sit with my lunch, in the crypt of St Martin's. England had been careless with its dead. Village churchyards were so overstocked that graves overlapped, spreading into putrid waste pits of human refuse. In the cities, crypts were at bursting point, coffins stacked in heaps, the atmosphere foul and fetid. There were fears of diseases rising like ghosts to claw the living into the bosom of the dead. For safety's sake it was decreed that the most offensive crypts should be sealed. From a notice in *The Times* in January 1859, Frank learned that one of these was St Martin-in-the-Fields. He was appalled. Somewhere among the mouldering stacks lay the body of the great John Hunter. It was intolerable that a man so honoured in life could be so dishonoured in death. Church officials confessed the shameful truth: they had never heard of John Hunter, and they had no idea where his body lay. Frank was adamant: Hunter had to be saved. The coffin must be found before the crypt was sealed, then reburied in the only place fit to receive it – Westminster Abbey. A painstaking search through the Register of Burials led him eventually to a cryptic entry dated October 1793, written in a 'crabbed and careless hand' and with a plethora of codes which, after some inspired cipher-cracking, enabled him to work out that Hunter had been taken to Vault No. 3.*

The search began on 1 February. What confronted Frank in Vault No. 3 was a state of chaos which, as I stir my coffee in this civilised dining room, I find impossible to imagine. The coffins lay like books tossed into a box, at all angles one on top of another, some piled across the doorway. On 7 February Frank's diary recorded that he 'turned out about thirty coffins, many curious ones among them. The stink awful; rather faint towards the end of the business.' Assisted by the vestry surveyor R. K. Burstall, he searched fruitlessly for another fourteen days, inspecting the coffins one by one as workmen dragged them out. There were more than two hundred of them in the vault, but by the afternoon of 22 February only five still remained. The tension mounted. Three of the nameplates were visible and none

---

* The vaults were brick compartments set between the piers of the crypt.

Illustration from *Reynolds Miscellany*, 1868, which may depict the
discovery of Hunter's coffin. Frank's handwritten note describes it as
giving 'a good idea of my work'

of them was Hunter's. Only two to go. Frank did not describe
what he felt when the men began to pull them out. Despair, the
fear of failure, cannot have been far from his mind. What he did
record was his 'shriek of delight' when the lamplight fell on the
penultimate coffin. 'I discerned first the letter J, then the O, and
at last the whole word John.' And there it was, the letters on the
brass plate as sharp as the day they were cut. *John Hunter, Esq.,
died October 16, 1793, aged 64.** But even now there was a price
to be paid. The sickly smells from the coffins, Frank wrote,
'were truly overpowering and poisonous'. He confided to his
diary that he felt 'quite knocked down and prostrate; thought
I was going to have fever; shivering pains in the back, head-
ache, shortness of breath &c. . . . thought I was going to have
a serious attack'. The vestry surveyor Burstall fared worse. He
was so ill that he had to spend four months away from London.

* Hunter was actually 65. Born 13 February 1728, died 16 October 1793.

Frank got what he wanted. On 28 March John Hunter was reburied in Westminster Abbey, on the north side of the nave not far from Ben Jonson. Next day Frank was invited to meet the council of the College of Surgeons to discuss a memorial for his resurrected hero. The result was the marble statue by Henry Weekes which still stands in the college's entrance hall, and a scholarship for comparative anatomy. Frank himself received a silver medal from the Leeds School of Medicine. In the scrap-book is a draft of his thank-you letter: 'Dear Sir, I assure you I find the greatest difficulty in finding words which will sufficiently express my extreme gratitude. . .' Not so much difficulty, how-ever, that he couldn't stretch it out for four densely written pages.

After lunch I walk the mile from Trafalgar Square to Lincoln's Inn Fields. Outside the Royal College – a detail that would have delighted Frank – two flabby middle-aged men wearing boxing gloves are circling warily around each other, aiming wild swings at each other's ribs. Inside the college I climb the stairs to the Hunterian Museum, named after the man on whose collection it was founded. The cabinets are packed with skeletons, parts of skel-etons, bone assemblies and pale organs like dissected ghosts in jars. Most of the material now is of human origin, though in Hunter's time, as in Frank's, much of it was animal. Frank knew very well how this might seem to a layman: 'The Museum of the Royal College of Surgeons in Lincoln's Inn Fields is, I fear, for the most part regarded by the general public as consisting of a collection of all the horrors of nature, of deformities, monstrosities, and malfor-mations, both human and animal, so collected and arranged as to form an enlarged "Chamber of Horrors", rivalling that in Madame Tussaud's exhibition.' He saw it as a blueprint for all creation.

Examining [the exhibits] as we walk along, the idea strikes us to ask ourselves what general law governed all these various creatures when in life? – why were they created, and why did they live? The answer is simple and easy. The great law of nature, to which all living things must submit, is, 'Eat and be eaten'; and in conformity with this law are all animal machines contrived and made to work. The plant eats decayed vegetable and animal substances; the herbivorous animals eat the plants; the carnivorous eat the herbivorous; man, as lord of all, lays his claim of power upon and over all.

Frank himself gave many specimens to the museum, and he made countless references to it in his writing. During his snake period he invited readers to notice a couple of specimens that had so overestimated the capacity of their stomachs that they had burst themselves through overeating. One had a toad exploding from its belly; the other a rodent more than double its own size. I try to find them, but they seem to have been among the large portion of Hunter's collection that was destroyed by the Luftwaffe in 1941. Most of Frank's gifts were lost in the same raid. Only four now survive, and none is on public display. One of these is his father's vertebrae. There is also a small piece of elephant ivory and the pickled genitals of a male dolphin. *Then there is the chair*. This must be one of the most hideous pieces of furniture ever constructed, but it is also one of the most moving: not just a bit of carpentry but a homage from a great man to a greater. It stands on a table in a storeroom, not sat upon I would guess for 120 years. Frank made it himself from John Hunter's bedstead, which Richard Owen somehow acquired and gave to him as a birthday present. It was an inspired gift. To Frank, anything touched by Hunter was an object of reverence. The dissembled limbs and spars of the old four-poster lay for a while in his room, waiting 'until the *quasi*-Hunterian presence should indicate to me what to do with it'. One morning it came to him. 'I cried out, "Eureka! I know . . . I'll make it into a chair." ' Then and there he measured the timber, calculated the depth, height and weight of the chair he would make, and got busy with his saw. The result is a peculiar object of polished dark wood, almost heroic in its ugliness, like the bastard child of a throne and a commode. Its fluted hind legs tower above the seat like minarets, and a huge carved wooden lump perches at the end of each arm like a pumpkin-sized pawn. Frank added a brass plate confirming its provenance* and a miniature copy of Joshua Reynolds's portrait of Hunter was set into the back. I am not allowed to sit on it but it looks painfully uncomfortable. Frank, being short, could not have leaned back and rested his feet on the floor – his legs would have stuck out like a toddler's. The only recorded instance of the chair coming into contact with a human bottom was during its official inauguration by Frank's old friend Dr. Wadham, physician to St George's, who sat in it while presiding over an annual dinner. But it is sad to see it now hidden away. One wonders how long it will stand here, this labour of love, unseen, unregarded and as distant from the public mind as Frank himself.

I cannot leave the museum without paying my own homage to Hunter's most celebrated trophy. Charles Byrne's skeleton stands at the end of the central display corridor, impossible to miss. Though five inches short of the advertised eight feet, he is still unimaginably huge. I am by ordinary standards tall

---

* The words on the plate read: 'THIS CHAIR IS MADE FROM THE BEDSTEAD OF JOHN HUNTER, BORN FEBRUARY 14, 1728. DIED OCTOBER 16, 1793. BURIED AT ST. MARTIN'S-IN-THE-FIELDS, CHARING CROSS, RE-INTERRED WESTMINSTER ABBEY, 1859.'

Heroic discomfort: the chair Frank built from the timbers of John
Hunter's bed, kept in the storeroom of The Hunterian Museum

but I have to crane my neck to view the skull. The limb bones
seem impossibly thin, the feet impossibly huge – big toes curled
around as if they had been crammed into too-small shoes. The
ribcage would swallow an oil drum. In a cabinet nearby is a wax
bust moulded from the skull, the reconstructed face showing a
wide fleshy nose and jutting brow. The word 'poignant' goes into
my notes: Byrne may or may not have been a gentle giant, but he
was surely a vulnerable one. Frank's breezy practicality for once
makes me feel sad. I remember his description of the hidden
room with the copper, where Hunter so jealously guarded his
privacy and where the big doors 'were so arranged that no smell

should escape except up the chimney, and that there should be more room than at first sight seems possible, for steaming as well as boiling'. No wonder the big man wanted to be buried at sea.

Another monstrous item recorded by Frank in 1859 was a whale, apparently a rorqual, which had been killed by fishermen and beached at Gravesend. On 6 May, accompanied by a museum assistant, Dr James Murie, he went to size the creature up. Its length was fifty-six feet; its weight an estimated forty-five tons. Though discouraged by 'the master' (presumably the sea captain who caught the whale) from climbing on the carcass for fear of bursting it, they somehow clawed their way to the top. 'It was hazardous walking,' Frank wrote, 'as the skin had all become loose and very slippery from decomposition; and there was not a little danger, as the tide was running down fast on all sides of this gigantic mass of flesh, which felt under the feet like a mountain of highly oiled India-rubber.' He noted in a subsequent account that local men had stripped some of the skin to make boots. Three days later he and Murie were back, unfazed by what must have been an overpowering stench, to dissect the monster and take home a lump of flesh for Frank to sample. This was one of the rare occasions when he bit off more than he could chew. Decomposing whale is not like well-hung pheasant. When a small whale was stranded near my home on the north Norfolk coast, the handkerchief-to-nose zone extended to more than a mile. Surprising only himself, Frank found the meat too high even when boiled with charcoal.

Next day he was at Hampton Court Palace, passing his naturalist's eye over Raphael's *The Miraculous Draught of Fishes*, in which he was able to identify a skate, a garpike, a dogfish and a haddock. As we shall see, he was an exacting art critic who reserved his praise for biological correctness. Two weeks later he accepted an invitation from the publisher of *The Field*, John Crockford, to answer readers' letters about natural history. His journalistic output beggars belief. He had no typewriter (the first commercial models did not appear until the 1870s), so every word had to be written in his somewhat inelegant hand. Even with the Internet, electronic keyboards and high-speed transport, few modern writers could match his total for the year of sixty-three pieces, let alone do so while holding down a commission in the army.

For Frank, the most significant event of 1859 was a dinner held a few months earlier on 21 January at the London Tavern in Bishopsgate. It was a date which he anticipated would mark the beginning of an 'epoch in natural history'. What would make it so was the menu. The eagerly awaited centrepiece of the feast was a haunch of eland, *Taurotragus oryx*, a large antelope from the plains of eastern and southern Africa. There to savour it were all the high priests of natural science. Frank described the occasion with his usual portentous enthusiasm:

> The savoury smell of the roasted beast seemed to have pervaded the naturalist world, for a goodly company were assembled, all eager for the experiment. At the head of the table sat Professor Owen himself, his scalpel turned into a carving knife, and his gustatory apparatus in full working order. It was, indeed, a zoological dinner to which each of the four points of the compass had sent its contribution. We had a large pike from the East; American partridges shot but a few days ago in the dense woods of the Transatlantic West; a wild goose, probably a young bean goose, from the West . . . The gastronomic trial over, we next enjoyed an intellectual treat in hearing from the professor his

satisfaction at having been present at a new epoch in natural history. He put forth the benefits which would accrue to us by naturalising animals from foreign parts, animals good for food as well as ornamental in the parks.

This was the point. It was not just to enjoy exotic dishes that the great men had assembled. It was to consider the possibility that the diets of ordinary men and women could be transformed by new kinds of meat. Owen wrote to *The Times* proposing that eland should be 'acclimatised' and bred for the English table. Frank was so impressed by this idea that he resolved to establish an 'Acclimatisation Society', dedicated to finding new sources of protein. His model for this was the French Societé d'Acclimatisation, which had been founded in Paris in 1854. It had more than two thousand members, enjoyed royal and imperial patronage and had a 33-acre garden in which to test its theories. Prizes were given for successful introductions of palatable animals, birds, fish, insects and plants. Frank was determined to do likewise. He cited Australia as the perfect example of a country transformed by imported animals and plants – a process in which Frank himself would come to play an important part – and reflected that only four new species had been added to England's livestock since the beginning of the Christian era: the turkey (1524), the musk-duck (1650), gold-pheasant (1725) and silver-pheasant (1740). 'Musk-duck' is a bit of a puzzle. The true musk duck, *Biziura lobata*, is a native of southern Australia, where it is not much shot at because people don't like the way it tastes. It is likely that he meant the Muscovy duck, bred for eating in Britain and descended from wild ancestors in Central and South America. The 'gold-pheasant' was presumably the golden or Chinese pheasant, *Chrysolophus pictus*, an ornamental species from eastern Asia which now breeds in small numbers in East Anglia. The 'silver-pheasant', *Lophura nycthemera*, is a forest species of South East Asia. If it ever did find its way to the English table, it certainly doesn't now. Frank might better have mentioned the guinea fowl, which was introduced in 1550 and is still enjoyed as a gamier alternative to chicken. What clinched the argument was the turkey. This alone, he said, 'is an answer to the question of the sceptic, who believes we have the best of everything; and, if he be a gastronomer, I appeal to that love of good feeding, which we all have more or less, and ask him, if it were not for the acclimatisation, which took place in 1524, what would he have for dinner at Christmas to face his roast beef?'

Bompas rightly observed that Frank's priority was to make science *practical*. 'To find out a new kind of food, or to multiply an old one, was to do a practical good to a hungry people; and to this end he henceforward devoted his chief energies.' The first meeting of the Acclimatisation Society was held on 26 June 1860. The same year saw the publication of a second volume of *Curiosities of Natural History* (known as the Second Series), which copied the first in its wide range of subjects and the good-natured acuity of Frank's remarks. Sometimes his observations were useful. He was keen, for example, to share his discovery that feverfew[*] kept moths away. Sometimes they were wrily amusing, almost self-mocking, and never more so than in his tragicomic quest for the mysterious *Podotherium*. This had begun on a London pavement where, outlined in snow and mud, he had noticed some very peculiar footprints. They

[*] Feverfew, *Tanacetum parthenium*, is a medicinal herb once prescribed to reduce fever (hence its name), now somewhat optimistically used to ward off migraine headaches.

were obviously human, but not like those of any normal man. The impressions were of a hobnailed boot, but the toe was two feet from the heel with nothing in between, 'so that the animal could not have had a sole to its foot'. Weirder still, the churn of mud showed that it walked *backwards*, and always along the edge of the pavement. Frank gave the strange creature a name – *Podotherium*, or 'beast with a foot' – and deduced that it lived somewhere among the crowded lanes of St Giles. After days of searching, he finally caught up with it in the Strand. *Podotherium* was only too human, a 'poor old cripple' who had lost both feet and walked on his knees. To make himself comfortable he had cut a pair of hobnailed boots in half, then strapped the heels to his knees and the toes to his lower legs. So the mystery was solved. Moving heel first gave the impression of someone walking backwards. The peripatetic nature of his trade, selling tin funnels, explained why he hugged the pavement's edge. It all served to underpin Frank's argument, that 'the observation of common things is necessary to the elucidation of unusual natural appearances'.

He was always drawn to people – fishermen, bird-catchers, gamekeepers – who could help him improve his knowledge. Gamekeepers fascinated and appalled him. *Fascinated* because the best ones could show him glimpses of rare or reclusive animals. *Appalled* by the mendacity of the rogues. To Frank, the most reliable inventory of any district's wildlife was the 'gamekeeper's museum', the nailed-up heads of 'vermin' – cats, hawks, owls, hedgehogs, stoats, weasels – culled for the sake of protecting game. A big collection might prove either the dedication of a skilled ecological cleanser or the rascality of a cheat. Frank noted the disgust of a landowner whose 'good-for-nothing gamekeeper had bought up all the cats in the neighbouring town, cut off their heads, and nailed them up, as trophies of veritable captures in the neighbouring woods; and how that same man had given a large order in an adjacent county for jays and magpies in order to swell the ranks of feathered culprits, and to palm off his purchased specimens as a *bona fide* collection shot and trapped by himself.'

Gamekeepers in the twenty-first century still nail up their 'gibbets', though there are some legal restraints. Frank was among many at the time who thought the best antidote to egg-stealing birds was a dead rabbit laced with poison (in one case he knew of, a single spiked rabbit had accounted for twenty-one magpies and crows). But the law now forbids indiscriminate killing. Laying poisoned bait is not allowed, and killing birds of prey carries a jail sentence. It is unthinkable that anyone now could write approvingly, as Frank did, of the use of strychnine by colonists in Australia 'for killing the dingoes, or native wild dogs, also the eagles, hawks, and vermin of that wonderful continent'. Nor would anyone be quite so gung-ho about the slaughter of cats. At one 'gamekeeper's museum' he counted fifty-three cats' heads that 'stared hideously down upon the visitor'. Cats' habit of killing for fun made them the gamekeepers' most hated enemy. The only species allowed to behave like that was the one with the gun. Frank wondered how anyone could be certain that a rabbit or bird had been killed by a cat.

'Because,' answers the keeper, 'every animal has his own way of killing and eating his prey.' The cat always turns the skin *inside out*, leaving the same reversed like a glove. The weasel and stoat will eat the brain and nibble about the head, and suck the blood. The fox will always leave the legs and hinder parts of a hare or a rabbit; the dog tears his prey to pieces, and eats it 'anyhow – all over the place'; the crows and magpies always peck at the eyes before they touch any part of the body.

Frank wanted to make clear that he felt no hostility to cats. Indeed he rather liked them, and urged his readers to consult the Honourable Lady Cust's informative little book, *The Cat, Its History and Diseases*. He admired feline agility and the way cats could fall from great heights without hurting themselves. How was this possible?

Why does not . . . she get concussion of the brain, as a man or a dog would . . . ? If we take down one of our dried cats' heads off the keeper's museum wall, and break it up, we shall see that it has a regular partition wall projecting from its sides, a good way inwards, towards the centre, so as to prevent the brain from suffering concussion. This is, indeed, a beautiful contrivance, and shows an admirable internal structure, made in wonderful conformity with external form and nocturnal habits.

The one animal conspicuously absent from these death shows was the dog. Even in those days an exhibition of dogs' heads – each one identifiable to its former owner – would have been a provocation too far. This did not mean dogs were spared retribution, but the evidence was likely to be buried. 'I have heard a theory,' said Frank, 'that the reason why the gamekeeper generally can produce finer gooseberries, cabbages, &c. than his neighbours, is that his garden is well manured with defunct dogs buried all about it.'

Hedgehogs also interested him, though in Hampshire his search for an undamaged skull was thwarted by the gamekeepers' habit of stamping on their heads. He had better luck in

Oxfordshire, where a gypsy woman roasted some for him over a fire. She told him that hedgehogs were 'nicest at Michaelmas time, when they have been eating the crabs which fall from the hedges'. Some of them, she said, 'have yellow fat, and some white fat, and we calls 'em mutton and beef hedgehogs; and very nice eating they be, sir, when the fat is on 'em.' Sometimes the animals were actually farmed. If the gypsies found a nest too small for eating, then they would tether them by their hind legs and wait for them to fatten. The hedgehogs' own dietary habits were a matter of academic dispute. Did they, or did they not, eat snakes? Frank settled it by shutting a hedgehog and a grass snake overnight in a box. The answer next morning was an emphatic yes.

Increasingly Frank was turning towards the sea. At Brighton he pondered the strange new fad of sea-bathing, which had turned ancient orthodoxies on their head. The famous good health of Northampton, for example, had always been explained by its distance from the noxious fumes of the sea. But now Brighton had become 'the paradise of invalids', and chemists' shops were springing up like toadstools after rain. Frank deplored the visitors' lack of energy and intelligent interest. He scorned the inertia of 'ladies and gentlemen . . . lounging on the piers, reading shilling novels and tales of love and murder, surrounded on every side by objects which, if they would only use their eyes, would afford endless amusement and instruction'. He showed them how it should be done. On the beach he measured the steady gradation of the shingle from 'the largest stones, ready sorted to the hand of the wall-maker', right down to the sand in the sea. Mixed into the shingle he found 'many interesting things' – fragments of ginger beer and wine bottles, tobacco pipes, shoes, knives, hard slag ('whence we might infer the presence of gas-works'), combs, hairbrushes and 'a ball, as large as a good-sized turnip, and quite as round, composed entirely of human hair'. Where had it all come from? The answer lay in the authorities' somewhat unsophisticated attempt to armour the coast against erosion. At the eastern edge of the town, rubbish carts had been discharging their loads straight over the cliff.

Frank gave a great deal of attention to a tiny shellfish called 'pholas' (*Pholas dactylus*, now better known as the common piddock) which caused grievous harm to ships and piers by boring into the wood and stone. It impressed him that Brighton's chain-pier had been piddock-proofed by thousands of nails hammered into the wood. Ever on the lookout for new ideas, he advised householders to follow the example of the pier's architect, Samuel Brown, and defend their homes in exactly the same way against 'another species of boring animal . . . the housebreaker'. As the modus operandi of this two-legged pest was to drill through doors or shutters, the solution was to drive broadheaded nails into the wood from inside. There is no record of how many acted on this excellent advice. After the piddock, Frank moved on to study the tenacity of mussels, the toxicity of jellyfish, the variety of seaweed (Frank preferred to call them 'sea plants'), cuttlefish ink and the grasping power of squid. Baring his arm to a squid was not an experience he wanted to repeat. 'The feeling is that of a hundred tiny air-pumps applied all at once, and little red marks are left on the skin where the suckers were applied; and when they were all fast, the animal could hardly be got off again. The sensation of being held by a "man-sucker" is anything but agreeable. The feeling of being held fast by a (literally) cold-blooded, soulless, pitiless and voracious sea-monster, almost makes one's blood run cold.' He recalled the gigantic specimens of 'cuttle-fish' (he should have said 'squid'), with fifty-foot arms,* which had been recorded in the Indian Ocean, and decided this must be the origin of the legendary sea-monster, the Kraken.

At Folkestone it amused him that the fishermen's name for dogfish – in Frank's own opinion the 'vagabond curs of the ocean' – was *Folkestone beef*. He was, as always, an acute observer. 'Ten minutes after the arrival of the boats the small fish-dealers may be seen cutting off their heads, tails and fins, and splitting them into halves; they are then salted and hung out to dry, and taste, when broiled, "like veal chops". They are eaten by the poorer class, "as a relish for breakfast".' The heads and intestines made crab bait. The livers were 'coiled and boiled for oil for the boats in winter'. Frank discovered that the heads were full of gelatine, but couldn't persuade the fishermen to boil them for soup like the Chinese, to whom gelatinous shark's fin soup was a delicacy. 'An Englishman is naturally a bad cook, and soup-making of any kind is not his forte.'

His Creationist faith led him always to seek perfection in design, and he found it in the dogfish. He pushed one head first into the sand and marvelled at how well it suited to its God-given purpose of rooting out flatfish. 'Man, curiously enough, copies its shape exactly in the instrument used for paring turf.' Only now, looking back from the chuck-away twenty-first century, do we see how thriftily the Victorians squeezed every last scrap of usefulness from whatever they could lay their hands on. The huss, also known as spiny dogfish or the 'rock salmon' of the fish-and-chip trade, had peculiarly rough skin which made very

---

* This is likely to have been an exaggeration. The maximum length of a giant squid, *Architeuthis*, is actually forty-three feet, measured from the posterior fin to the tip of the tentacles. The colossal squid, *Mesonychoteuthis hamiltoni*, is thought to reach forty-six feet.

serviceable sandpaper. 'It is a curious sight', Frank observed, 'to see the fishmonger's boy working away at his board with the head of a huss as a scrubbing brush.'

He seems to have developed a peculiar (if temporary) aversion to whelks. Being 'horribly carnivorous', as Frank put it, they were no trouble to catch. You could scoop them up with an open basket baited with stinking fish. Their instincts were unpleasantly rat-like. 'I have heard terrible stories of these whelks devouring human bodies found floating out at sea from wrecks – one in particular, where the flesh of the face and hands of a sailor had been quite eaten off, besides other instances too horrible to repeat. I always shudder when I see people eating whelks.' This sentimental nicety might be taken with a pinch of salt. We soon find Frank enjoying whelks at London market stalls, and he had no qualms about the equally unsettling (but infinitely more delicious) lobster. He reprised a story of his father's, about a shipwreck off the Isle of Portland. The people of Weymouth afterwards refused to eat locally caught lobsters and prawns, which they believed had fattened on the drowned sailors. The catch instead was packed off to London, 'for the benefit of those who did not know [its] history'. In lieu of human carrion, Frank reckoned the best bait for prawns was a fresh sheep's head.

He could be strangely forgiving of sharp practice among fishmongers. While he despised those who weighted their eels with sand, he admired the ingenious use of 'pouters', or, in fishermen's language, 'stink-alives'.* By skilful trimming, a pouter could be made indistinguishable from a whiting, thus inflating its value from a penny to sixpence. It was forgivable in Frank's

---

* *Trisopterus luscus*, now known as the pout or pouting, one of the lesser members of the cod family. Like the whiting it is not highly favoured, though fillets are now sold in some supermarkets.

opinion because both were 'as good food as the other when nicely sent to the table'. There were many worse things for him to worry about. He could see all too clearly the looming problems of pollution and overfishing, and complained about the waste of undersized flatfish caught in trawls. 'This wholesale slaughter in time will tell on the number of fish in the trawling ground, which is but a limited space after all.' This was an issue which, near the end of his life, would come back to bite him. He despaired also that eel boats could no longer work on the Thames beyond Gravesend. 'Even eels, that will live almost anywhere, can't exist in its waters, except at a great distance from the mouths of drains and other abominations, which pour into it a poison reminding us of the plagues of Egypt.'

In 1860 even Frank himself cannot have realised how propitious these thoughts would turn out to be. Most auspicious of all, on a visit to Scotland on 21 August, he caught his first salmon. It was, he said, the happiest day of his life.

CHAPTER EIGHT

Notice to Sick Porpoises

Anyone now trying to pigeonhole Frank would have a hard job of it. What was he? Naturalist? Journalist? Author? Lecturer? Military surgeon? Secretary to the Acclimatisation Society? Somehow he managed to be all these at once. And now there was something else. His friend Abraham Dee Bartlett had been appointed superintendent of London Zoo* and Frank had agreed to be his honorary consulting surgeon. Regent's Park for Frank was heaven on earth. He doctored live animals and dissected dead ones, while writing it all down and trying new things to eat. If he ever had an idle moment, he did not have time to record it. For a biographer it is like being hitched to a bolting horse.

His interest was unconfinable. Enthusiasm for the lower orders – rats, moles, pouters, piddocks – did not mean he was blind to star quality. Nothing excited him more than a lion, or moved him more profoundly than the moment in Psalm 104 when, as he put it, the 'deep bass notes of the organ and the swell of human voices proclaimed that "The lions, roaring after their prey, do seek their meat from God" '. When Sir Edwin

* For the sake of clarity and convenience I use the modern term, 'London Zoo'. Frank always referred to 'the Zoological Gardens'.

Landseer's four bronze lions were installed around Nelson's Column in 1867, the two most involved visitors were Frank and the man who probably knew more about big cats than anyone else in England, Abraham Dee Bartlett. Together they prowled around the statues making notes and 'mental photographs', before dashing back to check them against the lions in the zoo. Apart from the lack of whiskers, which no reasonable critic could expect to see cast in bronze, Landseer's lions were perfect. 'The complicated flexures of the opening nostril', Frank noted, 'are true to the eighth of an inch.'

England's adoption of the lion as a national symbol struck him as an entirely proper reflection on the country's 'gigantic strength', its playfulness, kindness and liking for fair play. He might have added its appetite for red meat. Every lion at the zoo was fed between eight and twelve pounds of beef (or horse) a day, including the weight of bone and sometimes a medicinal sprinkling of sulphur. Frank loved the old patriarch, monarch of the Regent's Park pride, and felt mixed emotions when it died. His first task at the post-mortem was to establish the cause of an apparently healthy animal's sudden death. Verdict: 'congestion of the lungs, caused probably by the excessive cold'. But what excited him was not so much the pathology of death as the structures of life. He was struck first by the lion's massive forearm, a rock-hard assembly of muscle and wire-like tendons whose circumference he measured at one foot seven inches 'or nearly the size of an ordinary hat'. He peeled the skin from the tendons to see how they worked the claws, and found they moved 'with the ease of a greased rope in a well-worn pulley'. His attention fell next upon the whiskers. 'We trace one . . . to its root, and find that it terminates in a mass of highly-sensitive nervous substance . . . and these whiskers are therefore so many watchful sentries which take their posts on dark stormy nights, when the regular duty men (the eyes) are unable to keep watch and guard.' Turning to the tongue, he found that its elevated spots, or *papillae*, were one-sixteenth of an inch long, and arranged obliquely in rows like the teeth of a file. This roughness, he observed, was essential to the work of a natural carnivore, enabling it to rasp every last shred of meat from the bone. He advised 'young observers' to prove the point for themselves by getting hold of a cat's tongue, snipping off the meat with scissors and leaving the skin to dry, when 'the spines will be then well seen'. His one fault may have been his failure to understand that not all 'young observers' shared his own sense of adventure, but he held absolutely to his belief that the best teacher is *experience*, the hearsay of textbooks no substitute for observation.

Many science professionals now think the same, though they are less likely to share his liking for travelling menageries. Frank's good opinion of these rested partly on a fallacy. He had noticed that lions born at the zoo had a peculiar suscepti- bility to a defect of the mouth. What he described, though he did not use the term, was cleft palate, which meant the cubs could not suckle properly. The travelling lions did not suf- fer in the same way, which made him think the problem had something to do with the zoo's fixed cages. He was thinking about this at about the same time as Gregor Mendel was for- mulating his mathematical theory of genetics, but thirty years before Mendel's thinking was widely understood. He did con- sider the possibility of a hereditary cause, but Frank dismissed it because the abnormality was not limited to a particular breeding pair, and because the sufferers were not related. With hindsight, we can see that he was probably mistaken. It *is* possible in rare cases for cleft palate to be caused by external factors – genetically disruptive chemicals, for example – but there is little reason to suppose these affected lions at the zoo, and none at all to believe the problem had anything to do with their permanent address. The overwhelming likelihood is that the cause was genetic. If captive-bred, the lions were likely to have had a recent common ancestor. If wild-caught, they might all have descended from the same local population with a common weakness. Frank might not have been right, but it was typical of him to think it was a subject worth thinking about.

Travelling menageries in the nineteenth century were big business, and none bigger than Wombwell's Royal Menagerie, which kept three shows permanently on the road. In 1858 one of these visited Windsor, where Frank was impressed by the 'startling roar of a lion thundering down the avenues of the Long Walk'. The exotic caravanserai was made up of fifteen vans hauled by forty-five horses with a staff of thirty or forty keepers, hardy souls who slept with the animals in the vans. At the showground they were drawn up in two lines with the ticket office at one end and the elephants at the other. There were two of these – a male, Akbar Khan, and his mate, Abdalla, which were the only animals not to ride in the vans. They had to walk, though never exposed to public view. They plodded along, ten miles a day, like enormous headless tortoises inside a 27-foot bottomless van which exposed only their feet. If you wanted to see a whole elephant, then you would have to put your hand in your pocket. Other Wombwell crowd-pullers included lions and cubs, a hyena, a bear, an Indian rhinoceros and 'war camels from the Crimea', which had been bought to replace a dead giraffe.

In 1856 the Surrey Zoological Gardens at Kennington – source of the panther chops Frank cooked at Oxford – had been shut to make way for a new music hall, and the animals put up for auction. The sale attracted dealers from Liverpool, London and the Continent, showmen seeking live gewgaws for their penny-a-view booths, taxidermists fancying the eagles, Wombwell's Royal Menagerie, London Zoo and Frank Buckland. Altogether a crowd of three hundred trooped around the showground, following a man with a bell. Frank did not intend to buy. He just wanted to see what value the market would place on the Creator's handiwork. Bidding started slowly: eight shillings for a 'wax-bill and two cut-throat sparrows', ten shillings for an armadillo, a pound for a pair of flying squirrels. Jackals were knocked down at twenty-four shillings a pair, and a brace of porcupines at £8 15s. A pelican fetched an astonishing eighteen guineas, which, said Frank, 'makes me think seriously of speculating in pelicans'. Monkeys were giveaways. A rhesus went for twelve shillings but none of the others made more than ten. One bidder even asked for an organ to be thrown in. Ungulates did little better. A mixed bag of an Indian goat, a 'four-horned sheep' and an Indian sheep fetched just two guineas – at which price, Frank reckoned, they would have been cheap even as mutton. But he thought £27 for an ostrich and £9 for a nilgai (a large Asian antelope) was fair. Excitement rose as the lots grew bigger and fiercer. A tigress was sold for seventy-nine guineas ('not her value', said Frank disapprovingly) and a 'very fine lion' for two hundred guineas. After particularly keen bidding, an elephant went to a circus for 320 guineas. That is between £31,000 and £38,000 at today's values, depending on how you do the arithmetic. Towards the end, a male (£62) and female camel (£50) were snapped up by Wombwell's, and a solitary giraffe was 'bought in' for £250. This last came to a sad end. Its journey to Europe ended at the docks when the lifting tackle snapped and it broke its spine in the fall. The carcass fetched £25.

A decade later two giraffes perished in a fire at Regent's Park, prompting Frank to campaign for safer stabling. He found it 'very painful to see these two beautiful creatures but lately in the highest health and vigour, motionless, charred, and inanimate'. Not so painful, however, that he couldn't enjoy the roasted flesh. According to Bompas (who might have been present) it was white and resembled veal. The auction finished with five bears. Even the best of these fetched only six guineas, and the cheapest four pounds. Frank was not optimistic about their futures: 'Poor things! They were . . . probably also defunct soon after the sale; for they were all bought by an eminent hair-dresser in the City.' In the mid nineteenth century the nostrum of choice for men losing their hair, heavily advertised and aggressively sold by barbers, was bear's grease. According to Alfred Rosling Bennett in *London and Londoners in the Eighteen-Fifties and Sixties*, it was sold in china pots at 2s 6d an ounce. To encourage faith in the authenticity of their product, some barbers claimed to kill their own bears. The trickery was ingenious, and Frank enjoyed getting to the bottom of it. He focused on a barber's shop in Hampton Street, off the Walworth Road near the Elephant and Castle, whose trade was impressively brisk. To keep up with demand, the barber had to kill three bears every week. Each was kept until the dread day of its demise in a barred enclosure where the barber, wary of its teeth and claws, would feed it scraps of meat on a pointed stick. Local schoolchildren, less easily frightened (and impossible to hoodwink), fed it buns. Sometimes the animal was a grizzly-looking oldster, grey with age. Sometimes it was a youngster, glistening black. Each new arrival was heralded by a notice: *Another bear just arrived*. As the children very well knew, and as Frank soon realised, no bear was ever killed. The animal was always the same, its appearance altered either by flour and grease to turn it grey, or by hair-blacking to restore its youth. The gaff was blown finally by a bellowing fishmonger known as 'Leather-mouthed Jemmy'. On killing days Jemmy was paid to mimic the roars and squeals of a dying bear, which he could do with extraordinary accuracy. One day the barber made a fatal mistake by refusing Jemmy his five-shilling fee. True to his name, Leather-mouth broadcast the truth far and wide, and the disgraced barber had to shut up shop and sell his one and only bear to a rival across the street.

Frank also loved circuses. He especially admired lion-tamers such as the 'Lion King', Isaac Van Amburgh, and the 'Lion Conqueror', James Crockett, and wrote approvingly of their training methods. Kindness, not cruelty, was the secret of their craft. '[The] days of hot pokers and carters' whips are past, and a common jockey's riding-whip is the magic wand by means of which the lords of the forest are subjected to the will of a man whose physical strength to theirs is as that of a mouse when in the clutches of our domestic puss.' This did not mean the lions had lost their natural instincts. On 14 January 1861, a stable hand called Jarvey entered the wicket gate at Astley's Royal Amphitheatre in Westminster Bridge Road, Lambeth (really a circus), and began raking the sawdust. Moments later an awful cry was heard and Lion Conqueror Crockett came running. Frank described the scene:

> Two lions were on the stage free as the day, and playing with the garlands left by an actress; the big lioness was up in the Queen's box, with her paws on the front of it, gazing out as proudly as possible; and near the stable-door, and about six feet from the gates, lay the man 'Jarvey', the lion sitting over him, as a dog sits over a bone.

Wielding a pitchfork, Crocket fairly soon had his troupe back in their cage. But it was too late for poor Jarvey.* For Frank it was a chance not to be missed. He hurried to the theatre and made a thorough examination of the corpse. His self-appointed task, scientific to a fault, was to 'read' the pattern of wounds and work out exactly how the lion had grabbed the man and mauled him. It did not take long. 'I account for there being so many more wounds on the left side than on the right side by assuming that the lion (as is its habit) cuffed him first on the right side and caught and held him on the left, just as we see a kitten playing with a ball of worsted.' Jarvey's death did nothing to lessen Frank's admiration for the circus. He believed there was a 'faculty in the human mind' which enabled some people to form close bonds with animals. You either had it or you hadn't. Some people would shudder when their paths were crossed by a toad. Others could fearlessly manage ill-tempered horses or slavering dogs. It was because of this 'mysterious power' that Frank saw the lion-tamers as 'remarkable men'.

On 27 November 1860, Frank lectured to a distinguished audience at the Society of Arts, once again making the case for acclimatisation. Nature, he said, had blessed the planet with some 140,000 species of animal,† yet only forty-three regularly saw the

---

* Crockett himself died in 1865, aged thirty, at a circus in Cincinatti where he had been earning the fairly princely sum of £20 a week. Cause of death was recorded as 'heat prostration' after Crockett on a hot July day had taken part in a long circus parade wearing a tin helmet.

† It is unclear which groups of species were included in this list, but it must have been more than just mammals, birds and fish (or was a misprint). The overall number of catalogued species in the world is now over 1.2 million, with 86 per cent of land species and 91 per cent of marine species still remaining to be discovered. Scientists' best guess at the likely overall total is 8.7 million.

inside of a cooking pot. To illustrate the point, he led his audience on an imaginary tour of the zoo, naming all the birds and animals with edible flesh that had been bred there. If it could be done in the zoo, then why not also in fields and farmyards? He piqued the interest of lay audiences too. A 'leading journal', quoted by Bompas, praised his talk on the 'Curiosities of Nature' as 'one of the most interesting, amusing and instructive lectures that have ever been given in Oxford'. Even allowing for journalistic hyperbole, this was a pretty remarkable tribute from one of the world's great capitals of learning. Frank was applauded even before he began to speak, and applauded repeatedly throughout a characteristically witty two-hour scamper through his current enthusiasms. Subjects picked out by the critic included elephants, mammoths, lions, tigers, hyenas, rats, snakes, spiders, caterpillars, birds, fish, oysters and pearls, 'all of which were illustrated and made more palpable by skeletons, skulls, relics, and specimens'.

At Windsor his subject was teeth. 'From the nature of the teeth,' he told his audience, 'the geologist ascertains the complete structure of the animal.'* Exhibits this time included the 'skull of a Caffre [kaffir] who harassed the rear of our army, and was just on the point of shooting a friend of mine, when he shot the Caffre instead, and kindly presented me with his skull'. Next up was the tattooed skull of a native New Zealander. Frank invited people to admire the tattoo, which 'imitated the graining of the root of a tree', then explained the context. 'Soon after the discovery of these tribes by Captain Cook there was such a traffic in these skulls among the sailors that the natives used to kill each other for the purpose of selling the skull of the victim, for which from £8 to £9 was sometimes given. This traffic is

---

* I can testify to the truth of this. My previous book, *The Hunt for the Golden Mole*, was about a species known only from teeth found inside an owl pellet.

now happily abolished and the skull is a rare specimen.' He went on to compare human dentition with the elephant's: different diets meant different teeth working in different ways (he likened the elephant's mouth to a grinding mill). A description of a gorilla was followed by a tiger's skull, 'which had had a man in its mouth, and carried the token of retribution in a bullet-hole'. He told them about the death of poor Jarvey, and the mistake he had made in trying to run away. Then came this gem from his post-mortem at the zoo:

Anyone visiting the Zoological Gardens would be attracted by the roar of the lion. An examination of the larynx shows it to be a most perfect musical instrument, resembling in its anatomy a trombone. There is behind the larynx a vibrating machine, which produces the sound by which all other animals are awed: a roar, a spring, and a pat with his powerful paw, and they are dead. The lion, like the cat, has a beautiful bunch of whiskers, which it distends whenever it takes a spring, and the moment the end of a whisker touches the least obstruction, a telegraphic message goes to the brain, 'Keep to the right, or keep to the left', and the body follows. Our whiskers are not so useful, only ornamental.

This would have raised a laugh. Frank's own whiskers were impressive but hardly ornamental; aesthetically their nearest relative was a rook's nest. He rattled on through his collection of teeth – viper's, sperm whale's, rattlesnake's, fish's – and finished with his usual homily, a breezy scamper through all the gifts of Creation from the 'leaf of the hedgerow' to the 'Divine artillery of the awful thunder'. This was both a hymn to the Creator and a swipe at His denigrators. Considering the animal kingdom in all its variety, Frank said, shall we not be struck with the thought that these things were not made in vain, that there must be an Omniscient Mind, which in infinite power and wisdom cares for the well-being of all His creatures, small and great? Shall we not have weapons of the most irresistible kind, wherewith to combat infidelity and the theories of those men who imagine they can, with their limited faculties, by notions of chance or mere properties of matter, explain and account for the actions of the great Creator? Shall we not at once become sensible of that privilege, which we are so apt to use without due thankfulness to its Giver, who has allowed us, with our imperfect senses, to contemplate and consider His perfect and beautiful works?

This was barely a year after the epochal sensation of Charles Darwin's theory of evolution. The first edition of *On the Origin of Species* in 1859 had sold out almost overnight and set the two opposing scientific camps – those who believed that Nature was ordained and delivered by God, and those under the influence of Darwin – at each other's throats. Here, clearly defined, was the dividing line between the faiths: inherited certainties on one side, evolutionary chaos on the other. It was a line that Frank would never be able to cross. He could not accept that the advent of man was a random chance event involving monkeys. In this he was very far from alone, and for a while the Old Testament gave the evolutionists a decent fight. Significantly, Frank's loathing of Darwin was shared by his friend and mentor Richard Owen. By a strange quirk of timing, Darwin's bombshell was swiftly but coincidentally followed by a decision to separate the British Museum into two distinct parts. Natural history, the 'works of God', was to be hived off from books, manuscripts and antiquities, 'the works of man'. When the new Natural History Museum in Kensington eventually opened its doors in 1881,

Genesis would still have the advantage. The eminent naturalist appointed to be its first superintendent was none other than Professor Owen, who conceived and designed his museum less as a scientific exposition than as a sermon encased in glass. It was a reverent hymn to the Creator, a zoological harvest festival in which evolutionary science had no voice. This is important, and we do Frank an injustice if we fail to take account of it. Though he may now look like a blinkered reactionary, he was not out of step with the century in which he lived.

Darwinism to Frank, and to those who thought like him, was no more plausible than witchcraft or primitive superstition, to which he was also vehemently opposed. It was widely believed in India, for example, that the best way to ensure the death of an enemy was to chop up the whiskers of a tiger and slip them into his dinner. Tiger-claw necklaces were prescribed to save children from the 'evil eye', and tiger fat to cure rheumatism. Frank was dismayed that nonsense of this sort could survive in nineteenth-century London, where an anxious mother only recently had asked Abraham Dee Bartlett for some hairs from the back of a lion to cure her child of fits. Worse, in 1862, a reputable London publisher, Longmans, produced a book* claiming that snake and lizard faeces would cure consumption (tuberculosis). 'The only person who benefited by the new kind of medicine', Frank wrily observed, 'was probably the keeper of the reptile house at the Zoological Gardens, the "excreta" being his perquisite.'

An engraving of Frank in one his later books shows him knee-deep in water, bottle-feeding a porpoise. It illustrates the tragic events of 27 November 1862. On that day he was at Regent's Park Barracks in Albany Street when a message came from

---

* *An Enquiry into the Medicinal Value of the Excreta of Reptiles* by John Hastings, MD.

Abraham Dee Bartlett: a live porpoise had been delivered! Frank ran to the zoo and found the animal exhausted and struggling for breath in a tank of seawater. He jumped in, held the porpoise up and poured sal volatile down its throat. The treatment seemed to work: its breathing quickened from a laboured eight to a healthier twelve respirations per minute. Two hours later a glass of brandy followed the sal volatile, and the rate went up to thirteen.* Frank now saw that the water was stained with blood from a wound in the animal's tail. He staunched the flow with salt, but then noticed that the bones of the left fin were broken. The animal evidently had suffered some rough handling.

It is important to remember that zookeeping in the nineteenth century was not the well-established, science-based specialism that it is in the twenty-first. Bartlett and Frank were pioneers doing their best to shine light in the dark. Animals brought to Regent's Park were not captive-bred specimens of known provenance. They were strangers whose diets, habits and health had to be established by trial and error. In this case, when the porpoise failed to improve, the two friends transferred it to the seal pond, where it perked up sufficiently to try a couple of swims, back and forth across the water. Each time it banged its nose against the edge, convincing Frank that it was 'very blind and stupid'. But he took full advantage of the opportunity for research. He might have been the first man in history to take the temperature of a porpoise's breath, which he said felt warm against his hand, 'like a jet of steam from a kettle'. He recorded fifty-three degrees Fahrenheit. But he did not expect the porpoise to survive, and it didn't. What he found post-mortem was horrifying even by

---

* This might not have been the good sign Frank thought it was. The average breathing rate of a relaxed porpoise is approximately four times a minute. In humans, which are of similar weight, the average is seventeen or eighteen.

Frank doses London Zoo's wounded porpoise with sal volatile. It was followed shortly afterwards by a glass of brandy

the standards of those rough-and-ready times. The porpoise had been caught by some Brighton fishermen who had first hit it on the head and then attacked its eyes. *Its blindness was real.* The cornea had been ripped by some crude pointed instrument such as a nail or piece of wire, and the lenses had gone. Frank learned afterwards that fishermen habitually put out the eyes of animals they thought might damage their nets. He deplored this as un-English, the result of poor education: 'If they were spoken to quietly and properly on the subject, I am sure these poor but honest fellows would leave off this horrid, cruel custom.'

This sad failure did not deflect Bartlett and Frank from their search for a live porpoise. To hurry things along, Frank wrote to friends near the sea and put notices in 'public journals' appealing

for a healthy specimen. A Mr Briscoe of Southsea put up a sign
on the beach:

NOTICE TO SICK PORPOISES
If Visiting this Beach,
their carriage to London will be paid.
A DOCTOR will be in attendance, and MEDICINE,
in the shape of
No end of Grog, will be found.
Please land early.
Apply to
FRANK T BUCKLAND
Regent's Park Barracks.   2nd Life Guards

The call was answered on 14 March the following year. A
telegram from a Mr Dutton of Eastbourne begged to advise
Frank that a porpoise awaited him on Brighton beach. Alas . . .
It seemed that some fishermen, inspired by the promised £2
reward, had caught the animal in a seine net and parked it overnight in a boat filled with seawater. When Frank arrived in the
morning it was dead, apparently the victim of foul play. As Frank
explained: 'The men thought some mischievous person had
killed him by placing a finger in the blow-hole.'

The saga continued. In October a Mr James Martin of Wigtoft,
near Boston in Lincolnshire, wrapped a live porpoise in wet grass
and blankets and put it on the train to London. Despite being
nearly eight hours out of water, it arrived safely and was slipped
into a pond with a sturgeon. According to Frank, the sturgeon
'seemed terribly jealous of the porpoise being put into *his* pond,
and swam about the bottom, round and round, looking as savage
as a fish *can* look'. The porpoise itself seemed perfectly happy,
and survived until the very day Frank announced its arrival in

The Times. Among the disappointed multitudes who flocked in the vain hope of seeing it was William Makepeace Thackeray, who celebrated the non-event in *Punch* with an 'Elegy on the Porpoise' as written by the sturgeon. It began:

> Dead is he? Yes, and wasn't I glad when they carried away
>   his corpus;
> A great, black, oily, wallopping, plunging, ponderous
>   porpus.
> What call had Mr. Frank Buckland, which I don't deny his
>   kindness,
> To take and shove into my basin a porpoise troubled by
>   blindness?
> I think it was like his impudence, and p'raps a little beyond,
> To poke a blundering brute like that in a gentlefish's private
>   pond.

The satire continued for another thirty-six lines.[*] It had a good effect, however, for more and more people now were throwing themselves into the challenge of finding a porpoise for the zoo. First came a telegram from Lancashire promising something even more exciting: a 'young, live, spouting whale'. Off they went, Frank and Bartlett together, 'dashing away northwards in the express'. Two hundred and twenty-eight miles later, they arrived at the door of a Blackpool bathhouse where a showman was drumming up a crowd. 'Walk up, walk up! Only twopence to see the monster of the deep!' Some monster! It was not a whale at all, but '*a poor little baby porpoise*, about two and a half feet long, floating at the top of the water'. Once again they came home crestfallen.

---

[*] This must have been one of the last things Thackeray wrote. He died on Christmas Eve of the same year, 1863.

Then on 22 November another telegram arrived. The sender, a Mr John Minter of Folkestone, had got hold of a live porpoise from a fisherman who had caught it in a sprat net. Frank rushed immediately to Folkestone, where he was met by Minter and a friend, W. Earnshaw, who had built a tank and filled it with seawater. The animal (which the fishermen persisted in calling a *fish*) was a fine, healthy specimen, though it was bleeding from a wound on its jaw. A pennyworth of stick-caustic soon fixed that; then the only question was how to get it back to London by train. It couldn't travel in Earnshaw's tank: the rocking of the train would splash water down its blowhole and choke it. After a few moments' thought, Frank instructed a carpenter to knock up a rough, coffin-like box lined with wet blankets. The Folkestone stationmaster kindly lent a guard's van, and a great palaver then ensued. A few minutes before the train was ready to leave, a party of fishermen led by Frank and the zookeeper Tennant, who had come along to help, hefted the porpoise from the tank into the box. Porpoise, coffin, keeper and Frank were then bundled aboard and off they went. They had not gone very far before the porpoise began to show signs of distress. The membranes around its blowhole were drying out so that it wouldn't close properly and the animal couldn't breathe. Frank spent the rest of the journey sponging the animal's nose, 'so anxious was I that he should live'. But that was not the only setback. Frank wanted the porpoise to get plenty of air, so he opened all the windows and shutters. As the train was travelling fast, the result was an icy hurricane – all very well until they entered a tunnel, when steam and smoke poured through the windows and into the blowhole, making the porpoise snort and sneeze so badly that Frank feared it would die. Its respiration rate shot up from nine a minute to a frantic fifteen. Miraculously, after a two-and-a-half-hour dash, the train rolled into London Bridge Station with a live porpoise still on board. Waiting for them was a cart with a fleet-footed horse, which careered through the streets 'like a fire-engine going to a fire'. A minor crisis occurred when Frank, while lighting his cigar, dropped a flaming match onto the creature's back, which did not please it. At the zoo they galloped straight to the reservoir, where Frank and Bartlett slid the porpoise tail first into the water. They sat for a while in the darkness, flashing a policeman's lantern to reassure themselves that their trophy was still alive, then went to bed 'quite tired out'.

Next day they had to find some way of feeding the animal. Frank's first thought was to dangle a herring in front of it with a fishing rod. This was not a success. The porpoise took the fish and gnawed it, but was too weak to swallow. Frank next tried a smaller piece of fish, but the result was the same. A small live carp was offered, but freshwater fish were not to the porpoise's taste and it was ignored. Several more bits of bait followed, but none of them would the porpoise swallow. The rest of the story is best told in Frank's own words:

Upon consultation . . . we determined, that as the beast was too weak to swallow of its own accord, that we would help him; so I got down by a ladder into the reservoir, and, catching the porpoise by the fin as he passed, watched my opportunity, and pushed a herring with my hand right down into his stomach; he scored my hand with his teeth, but I did not care about that. For a minute or two after I had given him the herring he seemed better, but he very soon showed that his supper did not agree with him, for he began to flutter his tail and dance about at the top of the water.

After sundry efforts, he made a spring, spat up the herring, and then – ungrateful wretch! after all the trouble and

labour we had bestowed upon him, turned up his fins and
died right off.

All Frank got for his trouble was a salutary lesson and
another poetic tribute, this time from 'Edward Ryley Esq, of
Leatherhead', published in *The Field* and taking the form of a
song, 'Frank Buckland's Porpoise'. This was a parody of a popular (if somewhat tasteless) comic ballad called 'The King of
the Cannibal Islands', and had a lively chorus: *Fishery, fleshery,
fowlery, jig, / Rum-tum-toodlum, little or big, / For whether a moa,
a porpoise or pig / 'Tis all the same to Buckland.*

The early 1860s were crucial for Frank. On 12 July 1862, the
Acclimatisation Society held a dinner at the fashionable London
social club Almack's Assembly Rooms,* in King Street, St James's.
The menu kicked off with birds' nest soup ('very peculiar and not
disagreeable flavour', according to Frank), followed by a soup of
tripan, or bêche-de-mer, a kind of Chinese sea slug. These were
as hard as horses' hooves, but after day-long soaking and night-
long boiling by the 2nd Life Guards' regimental cook they were
rendered into something more like common garden slugs. Frank
chopped them up before anyone else could see them, and went
on furiously boiling and simmering. The result tasted 'something
between a bit of calf's head . . . and the contents of the glue-pot;
excellent for a hungry man, and doubtless exceedingly palatable
to John Chinaman'. Then came *nerfs de daim*, soup made from
boiled-up deer sinews ('good eating, but glue-like'). Entrées
included 'Kangaroo steamer', a stew made from 'choice portions'
of the eponymous marsupial. Their choiceness, however, was a

---

* Bompas referred to the venue as 'Willis's Rooms', but it was not known by
this name until 1871.

little dented by mischance: the tin in which they were delivered was loose-lidded, so the kangaroo had somewhat gone off ('but not bad for all that'). A 'Pepper Pot' of spiced meats from the West Indies was so popular that the waiters had to be restrained from filching it ('I was obliged to tap their fingers with a spoon'). Dishes of curried chicken and spiced sweetbreads both went down well, too. But this was only the beginning. Guests who hadn't already fallen off their chairs now faced a whole Chinese lamb ('capital eating, for his bones were in ten minutes picked as clean as if a flock of vultures had been at him'), kangaroo ham ('rather dry'), wild boar ham, Syrian pig, Canadian goose, a trio of birds from South America – guan, curassow and Honduras turkey – and 'leporines', part hare, part rabbit, supplied by Abraham Dee Bartlett. *Entremets* included Algerian sweet potatoes, seaweed jellies and *Diocorea batatas*, or Chinese yams, 'from the branch society in Guernsey'. Desserts included dried banana, guava jelly and, mysteriously, meat biscuits ('Made in Australia. Useful for Sportsmen and Travellers'), all washed down with various chicory, tea and coffee substitutes, with wines and liqueurs from Victoria, New South Wales, Queensland, Guadaloupe, Algeria, the Ionian Islands and Martinique.

By now the society had published its first annual report naming among its patrons three dukes, three marquesses, eight earls, seven viscounts, three knights and sundry gentlemen of commerce, law and science, including the Chief Justice, the president of the Royal College of Surgeons and Richard Owen. Its president was the Marquess of Breadalbane, who would be succeeded first by the Duke of Newcastle and then by the Prince of Wales. Annual membership was two guineas, life membership £10. The principal benefactor, Miss Burdett-Coutts,* had given

---

* The banking heiress Angela Burdett-Coutts, one of the wealthiest women in Britain, described in a London paper as 'a living fountain of charity'.

£500. The society's achievements so far had been modest: Canadian quails were thriving, Chinese yams were growing, new varieties of peas and beans had arrived from North Africa. The secretary (Frank) had been authorised to buy a flock of the 'very useful and curious' Chinese sheep, to plant another acre of Chinese yams, and to send an emissary to Prussia for a stock of 'Sander' (zander), a freshwater fish which is related to the perch but looks like a pike. There had also been talk of various species of exotic deer, kangaroo, capybara and beaver, and the Admiralty had ordered ships' captains to bring back potentially useful specimens from around the world. The society's establishment credentials could hardly have been more exalted. Even so, it was easily mocked. New thinking always is. In the scrapbook is a cutting from the *Gardener's Chronicle*, which some joker had managed to trick into accepting a spoof annual report.

In Birds a great success has been obtained by the Hon. Grantley Berkeley,* who has succeeded in producing a hybrid between his celebrated Pintail Drake and a Thames Rat; and the Council consider that this great success alone entitles them to the everlasting gratitude of their countrymen, as this hybrid, both from peculiarity of form and delicacy of flavour (which partakes strongly of the maternal parent), is entirely unique.

From the perspective of the twenty-first century the Acclimatisation Society looks more like an idea for a musical

---

* The Honourable George Charles Grantley Fitzhardinge Berkeley, an MP, writer and sportsman who once fought a duel over an unfavourable book review. Like Frank Buckland, he contributed articles to *The Field*, where he established a reputation for vehement but not always rational argumentativeness.

comedy than a workable food policy. But it intended no more than we have since achieved for ourselves, a necessary increase in the variety and availability of things to eat. In fact the most significant sentence in the annual report, the one that pointed most clearly to the future, was buried in Frank's third-person account of his own campaigning speeches: 'He also exhibited drawings of useful fish, and also pointed out the facility of the details of the much-neglected art of Pisciculture.'

This was the thought that would promote him from celebrity to national asset. Astonishingly while all this was going on, he was still serving in the army, where, said Bompas, he was 'the author of endless merriment'. At a church parade in 1862, the respectful murmurs of worship erupted into laughter when Frank suddenly appeared in the company of a dwarf and a seven-foot-eight French giant who had joined him for breakfast. But the military chapter in his life was almost over. Encouraged by his income from writing, and by the £2,650 left to him by his father, he was ready for a new start. In April 1863 he resigned his commission and moved into the former home of Charles Dickens's father-in-law, George Hogarth, in Albany Street,* just ten minutes' walk from the zoo.

Four months later, on 11 August, he married Hannah Papps. Given that he first met her in 1850 and Poor Little Physie was born in 1851, we may wonder why it took him so long to legitimise the relationship. Neither Bompas nor Frank himself had anything to say about this, and in the echoing silence we can only speculate. What we do know is that the young Hannah was a girl far beneath Frank in social class, and we may think she looks a lot like a youthful indiscretion. The fact that Frank delayed

---

* No. 156, later renumbered 37. The site is now occupied by the Royal College of Physicians.

the marriage until his parents were dead strengthens the
suspicion that they were never made aware of her, or of the birth
and death of Little Physie. And it was only after the marriage
that Bompas acknowledged, or perhaps realised the existence
of, the woman he carelessly called 'Henrietta Papes'. We might
wonder, too, why there were no more children. Was Hannah for
some reason unable to have a second child? Or did they never
again share a bed? There is no evidence that their lives together
were unharmonious – quite the contrary – but there is no escaping the impression of a man driven more by honour than by love.

CHAPTER NINE

Robinson Crusoe

No one who had the slightest acquaintance with Frank would have expected his marital home to resemble anything from the pages of Mrs Beeton. He recorded very little about his wife, but there are frequent references to the torments visited upon his cook. Mrs Beeton would have been of little help to this valiant woman in her struggles with elephant-trunk soup, roasted giraffe-neck or any of the other body parts that Frank brought home after post-mortems at the zoo. Few cooks then or at any other time would have had to put up with their employer honking at them with the voicebox of a goose while slaving over fried viper. Blissful were the days when he offered nothing more challenging than fish.

Fish increasingly were on his mind. Britain was surrounded by salt water and veined with fresh, and in those waters swam fish. It was what the world now calls a no-brainer. Never mind what the Acclimatisation Society might achieve with African antelopes, Syrian pigs or Australian marsupials. A dependable supply of affordable fish would transform the lives of ordinary families. In early 1861 Frank began seriously to think about how he could make it happen. Once again he looked to France. Like the Germans and Chinese, the French had long ago developed artificial hatching techniques to maximise the supply of freshwater fish. Frank felt that England should do the same. The method was almost childishly simple: an arrangement of fish eggs, gravel and running water, hardly less natural than nature itself. A box would be filled with fine gravel which had been sterilised by boiling. Fish eggs would be dropped into the gravel and a slow dribble of running water – enough to keep the eggs alive but not to wash them away – would be released from a cistern. This model stream bed could be enlarged with boxes placed one below another like the steps of a waterfall. Bompas described how the miracle then unfolded:

> In the eggs, which resemble small whitish semi-opaque beads, presently two black specks appear, which develop into eyes, and in due time the egg bursts and the embryo fish uncurls, half an inch to one inch in length, with large eyes, closed mouth, and a transparent pouch of nutriment hanging underneath the throat, to be gradually absorbed in about six to eight weeks, when the mouth opens, and feeding by that organ begins.

Frank began his experiments in May 1861, using perch eggs from 'Surley Hall' (an inn by the Thames, since demolished) which he hatched at home. His hatchery drew swarms of fascinated visitors, a must-see for anyone interested in freshwater fish, and he soon found himself catapulted onto the council of the British Fisheries Preservation Society and the committee of the Thames Angling Preservation Society. He did not disappoint. Like a magician he caused a sensation at a meeting of the Thames group by theatrically revealing, *hey presto!*, the baby perch he had just hatched. But perch were never going to be enough for him. Though they were valued by Victorian cooks (Mrs Beeton served them poached with parsley and melted butter), he quite literally had bigger fish to fry. On Christmas Day 1861, he planted his first salmon ova in hatching boxes at the home of a friend, Stephen Ponder,* by the Thames at Hampton. This was clearly a moment of some importance (why else would he give up Christmas Day?), but just *how* important would not become evident for another year or two. It was in 1863, after returning to civilian life, that Frank (in Bompas's words) 'threw himself with renewed ardour into the work he had now chosen for life'. For renewed ardour read 'manic energy'. The hatchery at Albany Street was only one small part of a burgeoning enterprise. The calls on Frank's time were endless. His advice was urgently sought by the many who were following his example, and who now needed ova for their hatcheries. Collecting fish eggs is hard work which has to be done during the spawning season. For trout, this meant wading in the near-freezing rivers of January and February. The hair oil with which Frank smothered his skin was an imaginative but probably useless attempt to ward off the cold (he later would advise the Channel swimmer Captain Matthew Webb to try porpoise grease).

He also corresponded regularly with the French government's Etablissement de Pisciculture at Huningue in Alsace, which sent him large numbers of salmon and trout ova. In January 1863 he built a hatchery in the window of *The Field*'s office in the Strand. It was a stroke of genius, a crowd-puller and a powerfully effective advertisement for his scheme. Victorians liked to see things for themselves. In the pre-filmic, pre-electronic age there was no other way for them to witness the wonders of the world. Hence the fortunes earned by showmen like Phineas T. Barnum

---

* Ponder was one of the most committed members of the Thames Angling Preservation Society, a fierce campaigner for the opening of weirs to allow free movement of fish.

and itinerant zookeepers like George Wombwell, and the flakier livings made by hustlers and freak merchants.

Frank's diary for February that year shows the fullness of his commitment. On the 7th he travelled to Gerrards Cross, where he gathered 'several thousand' trout eggs for *The Field*. On the 9th a batch of Rhine salmon hatched out in his bath. On the 12th he collected 'about 3000' salmon ova from Carshalton. On the 17th 3,000 salmon trout and 2,000 Great Lakes trout were delivered from Huningue. Salmon ova arrived from Galway on the 20th, and on the 21st Abraham Dee Bartlett and the angling writer H. Cholmondeley Pennell helped him harvest 'between three and four thousand' ova at St Mary Cray. On the 24th he gave a lecture on fish-hatching at the zoo ('a great success'), and on the 25th dissected salmon at the British Museum. The 26th brought a delivery of 3,000 Great Lakes trout eggs, 1,500 'Ombre Chevalier'* and 2,000 salmon trout. These were followed on the 28th by another 2,500 Great Lakes trout. But none of these things would have the enduring resonance of the entry for 3 March: *With Mr Youl by appointment, and examined the salmon ova he had packed in ice, which had been there forty-five days.* The ripples this caused would run all the way from Albany Street to the far side of the world.

Among those who admired the hatchery at *The Field* were Thomas Henry Huxley and William Henry Flower, conservator of the Hunterian Museum, who in 1884 would succeed Richard Owen at the Natural History Museum. Frank lectured at the Royal Society and the Royal Institution, and (odd as it may now seem) in July 1863 exhibited his hatching apparatus at the Islington Dog Show, where the Prince of Wales was among those who came to seek him out. Further demonstrations

---

* Probably a misspelling of *omble chevalier* – French for Arctic charr.

followed at the South Kensington Museum and at the Crystal Palace. According to Bompas, the 'grave members' of the Royal Institution were moved to 'laugh heartily at the racy humour' with which Frank made his points. This encouraged him to expand the lectures into a book, *Fish Hatching*, which he dedicated 'in the name of English progress' to M. J. Coumes, chief engineer of the Etablissement de Pisciculture. I must say the racy humour is elusive to this particular reader's eye – the tone might better be described as chummy – but if you want a step-by-step guide to hatching fish, then look no further. The book is a masterpiece of simplicity. There was no great mystery to solve. The basic principle of a fish hatchery, Frank explained, was to replicate as closely as possible the natural environment

Drawing from *The Field* of the Islington Dog Show, 1863, at which Frank's fish hatching apparatus was inspected by the Prince of Wales. Frank himself is depicted, second right

in which a species had established itself. It gave him yet another opportunity to delight in the perfection of design (a Darwinian would say *adaptation*) that fitted an animal to its environment. By 'design' he meant not just physical form but also the workings of the brain. To those still sceptical of animal 'intelligence' he gave a gentle chiding in a book of essays* edited by H. Cholmondeley Pennell. It was never wise, Frank thought, for a human to believe himself superior to another species in its element.

If we listen to a lecture from a learned professor upon the brains of animals, he will point out the human brain as being at the highest end of the scale, the brain of the fish at the lowest. Holding up the brain of a fish, beautifully prepared in spirits of wine, he will say: 'There, gentlemen, is an example of a badly-developed brain. The creature to which it belonged is proverbially dull and stupid.' Yet the next day, if we look over Richmond-bridge, we may behold the same learned but sportless professor puzzling his well-developed brain to catch the creature which but yesterday he was asserting had so little brains. The brain of the fish is quite sufficient to keep him off the professor's hook, angle he never so wisely.

While all this was going on, Frank and Hannah were settling into their new home. For Frank there could have been no more desirable an address in the whole of London than Albany Street. Whether proximity to his old billet at Regent's

---

* *Fishing Gossip; or Stray Leaves from the Note-books of Several Anglers*, edited by Pennell and dedicated to the late MP for Sunderland, Henry Fenwick, who had been an energetic campaigner for fishery reform.

Park Barracks mattered to him, I rather doubt. The important
thing was that Albany Street ran parallel to the Outer Circle
along the eastern edge of Regent's Park, within a lion's roar
of the zoo. If a camel sneezed, or a rhinoceros complained of
toothache, then he could be there in minutes. Bompas's
word 'unconventional' to describe the household falls well
short of the mark. The most complete account of this unique
ménage appeared several years later in *The World* newspaper's
'Celebrities at Home' series. On arriving at the house, the col-
umnist reported, the visitor would be ushered into what the
previous owner, George Hogarth, would have called his drawing room. For Frank and Hannah it was at first the 'Master's
room', and later the 'monkey room'. 'Darwin going backwards'
was how Frank explained it to the man from *The World* as the
monkeys tormented a young jaguar. This 'jolly little brute' had
arrived at no. 156 as the sickliest of babies from the zoo. Its
forelegs were paralysed and it was not expected to live. What
makes this story worth telling is that it was not Frank who
nursed it back to health and made a pet of it, but Hannah. The
writer described it nestling like a puppy in the folds of her
dress. It is one of the best indications we have that Hannah was
an active participant in her husband's work, and an important
part of the small-animals clinic which the house had effectively
become. Frank's concession to domesticity was limited to the
rule – rarely observed – that living animals should be barred
from the dining room. 'It is held,' said the writer, 'so to speak,
at the sword's point, against the incursions of animals from
the neighbouring jungle.' Dead ones were a different matter.
'It is regarded as Poets' Corner for the great, while the bodies
of the less distinguished are consigned to honourable burial in
the back garden.' That Frank was not conventionally house-
proud is a truth hardly worth stating. 'Mr Buckland's chief

domestic grievance', observed *The World*, 'is the duster, which he regards as a mischievous invention of women.'

Each day the parcels van dropped off a supply of malodorous corpses for Frank to dissect. One day Hannah arrived home to find a cask with a gorilla* inside it. And the excitements were not confined to the house. According to *The World*, Frank kept a laughing jackass (or kookaburra, *Dacelo novaeguineae*) whose favourite trick was hailing taxis. Frank was proud of this bird – 'as fine a jackass as could be found within a hundred miles of St Paul's' – but frustrated by its refusal to live up to its name. He thought he detected a 'slight titter' when he gave it a live mouse, but that was as merry as it got. Inevitably one day it escaped and flew off into Regent's Park. 'One parting farewell only he gave me; the rascal actually stopped in his flight, and for the first and last time I heard his hearty laugh.' The bird was eventually caught by a man in Stanhope Street who found it asleep at the foot of his bed, and Frank then clipped its wings. Confined to barracks, it took to terrorising Frank's 'big Turkish wolf dog', Arslan†. Amid the exotica of 156 Albany Street, cats and dogs seem hardly worth mentioning. Arslan, however, was an exception, a notorious terrorist whose pleasure was slaughtering the neighbourhood lapdogs. His career ended when he sneaked through a window and snatched a pet from the arms of its terrified mistress. To restore calm, Frank exiled him to the guard ship on the Herne Bay oyster beds.

---

* This was not a unique occurrence. The first gorilla to reach London Zoo, in 1858, arrived in a barrel of spirits. A photograph of Abraham Dee Bartlett posing with it appeared in my previous book, *The Hunt for the Golden Mole*.
† Presumably an Anatolian Shepherd dog. Frank may have named Arslan for the 11th-century Turkish sultan and warlord Alp Arslan, whose name in Turkish means 'heroic Lion'. Or he might just have meant 'lion'. C S Lewis's Aslan, representing Jesus in the Narnia stories, derives from the same root.

All the more extraordinary, then, that such a monster could be frightened of a jackass. Frank plainly loved animals, and loved them unconditionally, but his pets were not just for amusement. They were for *observation*, and contributed much to his writing. He wanted to know: was the dog justified in its fear of a mere bird? Exactly how dangerous *was* the jackass? How did it tackle its prey? This was of particular interest to Frank, as in Australia the bird was known to kill snakes. For his experiment he had to make do with a frog, which he released onto the floor. Sensing danger, the frog hopped away as fast as it could, but the jackass was faster.

He caught him with his big bill, and made a motion as though he was going to swallow him forthwith, but, changing his mind, suddenly he hit the frog a most tremendous blow on the floor, and, in a minute or two, repeated the blow, knocking the frog about, as one sees Punch knock about the constable with his wooden staff. This was most interesting to observe, inasmuch as it proves the way in which this bird, in his native country, is enabled to kill a venomous snake without its being able to bite or kill him, for, being so shaken and knocked about, the snake has not time to turn round and bite.

This unshakeable faith in experiment, the reluctance to accept as fact something he had not seen for himself, was the explanation for much of Frank's supposed eccentricity. A friend who found him one day cooking a slice of kelt (an emaciated female salmon after spawning) tried to dissuade him from eating something so obviously disgusting. 'No doubt', said Frank, 'it is nasty enough, but how can I say so unless I have tried it?' Yet there were times when even Frank had to admit defeat. The same friend described him throwing in the towel against a peculiarly gross oyster, 'the size of a cheese-plate', though he was unwilling to abandon the experiment entirely. A dustman was offered a shilling to swallow the thing, but even this hardened connoisseur of filth couldn't get it down. Frank added a pot of porter to the bargain, which encouraged the dustman to swallow half the oyster before he too retired in disarray. The precise flavour of this oddly indelicate delicacy was 'locked unrecorded in the dustman's breast'.

The stream of humans through the house was no less exalted, spectacular or exotic than the wildlife. The *exalted* included a continuing procession of Britain's finest scientific minds. The *spectacular* included some of the most extreme variations that nature ever imposed upon the human frame – giants, dwarfs, conjoined twins. The *exotic* included anthropological headliners from across the globe. Among these was a band of New Zealand Maori chiefs led by Tomati Hapiromani Wharinaki. After a convivial dinner, Frank invited his guests to take a pipe or two with him in the monkey room. As was his custom, he sought to amuse them with a display of his most recent acquisitions. *Look here!* He flipped open a box and spilled three dozen slow-worms onto the floor. Bompas described what happened next:

> Instantaneously the guests were transformed; the garb of civilisation slipped off, and they returned to the wild untutored savage. With one frantic glance at the slow-worms on the floor, they uttered wild yells and straightaway fled. Downstairs the dining-room window was open, through this helter-skelter into the garden, like hounds breaking cover, and filling the air with a *tapage d'enfer*. Thence they spread over the neighbouring gardens, taking

fences like deer. Two of them seeing another open window, and at it a peaceable old lady at work, headed for it, dashed in, and with their tattooed faces and awful cries nearly were her death. By this time the whole parish was up, a hue and cry organised, recruits joined from the railings, and the fugitives were run safely to ground.

It turned out that the Maoris had a superstitious horror of the slow-worm, which they believed was *ngarara* – the incarnation of the power of evil. Another fact duly noted. Frank also took the New Zealanders to the zoo, where they were terrified by the elephant, delighted by a zebra (which they seemed to think was a tattooed horse), and offered to repay Frank's kindness by tattooing his face.

It was in the summer of 1863 that Frank first met George Butler, an elderly one-handed fisherman who worked out of Portsmouth. Butler might have been his name, but his weather-beaten appearance and makeshift vessel – 'the smallest boat that could contain a human frame' – meant he was known to all as 'Robinson Crusoe'. He had lost a hand thirty years earlier while serving on the frigate *Hind* off Cape Horn. 'I got him jammed between two water casks,' he told Frank, 'and the doctor took the slack of him off.' (Admirers of Patrick O'Brian's Aubrey–Maturin novels will notice the echo.) Crusoe's souvenir of the accident was a severed finger preserved in a bottle of rum. He could still feel the wart on the thumb of the missing hand, which he reckoned was the perfect thermometer.

Hungry for sea-lore, Frank bought Crusoe's entire catch of pout (it didn't amount to much) and examined the small miracle of his boat. In shape and size it was more like a sawn-off water butt than a boat, the planks so rotten that a good kick would have gone straight through. There were holes in the side which Crusoe had patched with bits of canvas, leather and tin. With only this thin membrane of hope between himself and the ocean floor, in storm or fog the old man rowed out every night, taking his chance with the tide and the steamers that might run him down without even noticing he was there. He told Frank he had once taught a kitten to swim and catch dogfish in its teeth. Its sudden disappearance one day was blamed on cat-nappers bound for the West Indies, or perhaps on some fur trader who fancied its fine black pelt. Cat-rustling by seamen was not uncommon. As Frank explained in the magazine *Once a Week*, there were two reasons for this. Marine insurance did not cover cargo damaged by rats; but if an aggrieved merchant could prove the vessel had sailed without a cat, then he would have a case for damages against the owner. 'Again,' wrote Frank, 'a ship found at sea with no living creature on board is considered a derelict, and is forfeited to the Admiralty, the finders, or the queen. It has often happened that . . . some domestic animal, a dog or canary bird, or most frequently a cat, has saved the vessel from being condemned as a derelict.' The old man yarned away about the water, of which he knew every ripple, the wrecks over which he fished,* the baits he used (boiled cabbage and onions were favourite) and the size of the catch (varying from zero to brimming boatfuls). A couple of days later, after a storm, Frank hired a boat and rowed out behind Crusoe in his tub. He was rewarded with enough fishermen's tales to fill an entire chapter (Frank did exactly that in *Curiosities of Natural History, Third Series*). It is striking how much tastes in fish have changed. For Crusoe, the

* These included HMS *Boyne*, flagship of Vice Admiral John Jervis, which caught fire and burned at Spithead on 1 May 1795.

best use for a sea bream was to bait his lines for pout when he had run out of lugworm. Bream is now a premium fish, comparable to the even pricier sea bass, while pout is often used to bait lobster pots or as cheap 'filler' to bulk out fish fingers, pies and other lowlifes from the supermarket freezer cabinet. Frank himself, despite his appetite for novelty, seems to have shared the conservatism of his contemporaries. He damned monkfish as 'not good eating', though the rough skin might be useful for sword handles and instrument cases. How things change! On the day I checked, monkfish fillet was offered online at £14.30 for 230g. Cod on the bone was £22.20 for 600g and wild turbot £23 for 280g. One century's crab bait is another century's gourmet treat.

Crusoe claimed never to sleep while at sea, though in the hour after midnight it would have made little difference if he did. That was when the fish slept, and no fisherman would get a bite until the clock struck one. This inspired a lively correspondence when Frank mentioned it in *The Field*. Do fish sleep? Well, *do*

Robinson Crusoe

they? Two aquarium owners were sure they did, though not all species were alike. One of Frank's correspondents, M.M., said his carp, tench and minnows all slept at various times but the goldfish were always awake. Another, F.Z.S., identified wrasse as the champions of somnolence. '[A]fter playing up and down for an hour or so, picking up their dinner of infusoria, they quietly retire to their corner, and, lying down on their sides, go (I think) comfortably off to sleep.' Writing later in Frank's own magazine, *Land and Water*, the writer-traveller Henry Hoare Methuen described bonito and tuna following his ship on a fast run to the Cape of Good Hope, day and night for hundreds of miles. 'All this time they could not have slept, and must, beside constant muscular exertion, have fixed their attention on the ship to keep her company.' It is a question that causes confusion even now. The US National Oceanic and Atmospheric Administration is careful to use quotation marks when it describes fish 'sleep' as an area of active research. Fish, it says, do not sleep in the way that mammals do, though most of them take a rest. They become still, and slow their metabolisms, but remain alive to danger. Some float while they rest; others wedge their bodies in mud or coral. Having no eyelids, they cannot close their eyes. These periods of suspended animation, says the NOAA, 'may perform the same restorative function as sleep does in people'. Even so, not every species seems to do this. Some swim continuously and show no sign of resting at all. Were Frank ever to be reincarnated as a fish, one feels he would be one of these.

There were few aspects of life in which he did not have an opinion, and sleep was no exception. He was not a man for bedtime cocoa. His preferred sleeping draught was raw onions. 'In my own case,' he wrote, 'it never fails. If I am much pressed with work, and feel that I am not disposed to sleep, I eat two or three small Onions, and the effect is magical.' He was not alone. Hilda Leyel, founder of the Society of Herbalists in 1927, attributed 'decided soporific power' to onions and prescribed onion gruel as 'a very helpful and safe remedy to take at night to induce sleep'. Those seeking confirmation on the Internet will find that the most commonly quoted authority for it is Frank himself. 'Because Buckland says so' was the clincher in many an argument of the 1860s and 70s, but it seems a strange context in which to find it now.

Frank had a profound respect for 'ordinary' men like Crusoe. He loved listening to him and wrote down hours of their conversation: on the mortal risk of hooking big congers from small boats; on convicts fished out of the harbour; on the Irish steamer which Crusoe once hallooed to warn it off Lump Rocks in a fog, and which rewarded him with stark ingratitude; on the perversity of 'foreign ladies' who defaced themselves with 'paint, putty and glue'; on the quality of soft-shelled lobsters, which, 'though the gentlefolks don't like 'em', were as good to eat as hard-shelled ones; on the power of lobsters' claws. Crusoe hooked a lobster, a valuable 'berried hen', as they were speaking. 'Spiteful as a mad dog', it sliced a pout clean in half. But lobsters in Crusoe's opinion were nowhere near as dangerous as crabs. He told Frank about a man on the Isle of Wight whose hand was clamped by a crab he had been trying to extricate from a hole, and which drowned him when the tide came in. Another old sailor, George Brewer, still fishing at the age of eighty-six, remembered fighting in 'the glorious 1st of June 1794', or the third Battle of Ushant. His yarns ranged from the hideous bloodbath of an eighteenth-century sea battle (more proof of the veracity of Patrick O'Brian) to the taste of fishermen's bait. When Frank saw 'Uncle', as Brewer was most often called, sluice out his 'lug-pot' with seawater before filling it with grog, he asked him if he had ever eaten a lugworm. 'Lord, ay, my son, to be sure I have. They are as bitter as soot, if you eats 'em raw, but they are as sweet as sugar if you cooks 'em.'

The best thing about this story is its ending. Frank revisited Crusoe with a friend and found him considerably spruced up: hair cut, beard shaved and wearing a most un-Crusoe-like clean shirt. The old sailor told them what had happened. One fine morning he had been hailed by a woman on a yacht who had read Frank's article about him in *The Field*. She invited him on board, gave him dinner and then presented him with a gleaming new boat. Crusoe was proud of his new craft, but only in the way a collector might be proud of an artwork. It stayed on the beach, and he refused to put to sea in anything but the leaking old tub that was certain soon to drown him. Frank realised the time had come for *force majeure*.

> After a deal of trouble, [my friend and I] got Robinson Crusoe to accept a price for [the old boat], about ten times her value. The bargain concluded, we hauled her up on the beach, and taking a run, my friend and I simultaneously gave her sides a good kick. This was quite enough: she fell to bits like an orange box; and with a big stone or two, we soon broke her into pieces so small that even Robinson Crusoe could not put her together again.

*

Frank himself was now sailing into rough water. The problem was the salmon hatchery he had established with Stephen Ponder at Hampton. He had expected this to be supported by the Acclimatisation Society, but it did not happen like that. The opponent (Frank himself would have said *enemy*) who blocked him was another of *The Field*'s regular contributors, Francis Francis, a Devonian sea captain's son who edited the paper's angling pages. Francis fancied himself as an expert on pisciculture and was determined to repel Frank's trespass

on what he regarded as his home turf. In July 1863 he wrote
to the Acclimatisation Society offering to set up a hatchery on
their behalf at his home in Twickenham. Caught off guard,
Frank protested in favour of his own scheme at Hampton.
But it was in vain. He succeeded only in having the decision
deferred for a week, and had to suffer the rare frustration of
losing an argument. It irked him that *The Field*'s editor in
chief, John Henry Walsh, claimed that Frank himself had
'highly approved' Francis's proposal. The magazine now
appeared to be at war with itself. Another of its writers, the
disputatious fox-hunting enthusiast Grantley Berkeley, who
was one of the Acclimatisation Society's vice presidents, now
decided to get involved. In fact he had already resigned from
the magazine after falling out with Walsh. The hatchery controversy was all the incitement he needed to wade back in, and
he wrote an incendiary letter to the *Dorset County Chronicle*.
'It has become evident to me,' he said, 'by the late substitution
of the *sub-editor of The Field*, a Mr Francis Francis, to the paid
position of manager of the fish-culture to the Society, instead
of that really clever gentleman Mr F. Buckland, that a clique
having a majority in and around *The Field* office has a great
deal too much to do with the Society ever to let its interests
stand on their own merits.' For this reason he was no longer
willing to serve as the society's vice president. But he hadn't
finished yet. The 'egregious absurdities' published in *The
Field*, he said, meant that it could no longer be taken seriously. His profound wish was that 'the sportsmen of England'
would come together and start a new, 'well-conditioned sport-
ing paper, with a gentleman at its head', that would sweep
The Field from its pedestal. We can't know whether or not
the editor he had in mind was the same 'really clever gentleman' he thought should run the Acclimatisation Society's

fish-hatching operation but in the light of what followed it seems likely that it was.

None of this served to narrow Frank's range of interests. He was a habitual (and one has to say somewhat prolix) writer of letters to *The Times*. On 20 December 1863, he had become strangely exercised by the dangers of poorly designed millinery. Looking into shop windows he had noticed a trend towards forward-pointing spun-glass 'peacocks' tails' attached like ships' figureheads to the prows of fashionable women's hats. This was to be deplored.

These plumes are highly ornamental and graceful to look at; but I would beg the gentlemen to warn the ladies of their families against wearing them, for these threads of glass are as thin as cobwebs, and, though apparently solid as a mass, break and snap off, falling into almost impalpable powder with the greatest ease.

Now, imagine the consequence; these *spiculae* are very likely to find their way *into the eyes* of the fair wearers; and I need not tell them what pain and discomfort they would necessarily cause. It would, moreover, be a difficult matter for the surgeon to see them, and when he had discovered them it would be delicate work to remove them from the sensitive membranes which line the eyelids.

On more familiar ground, Frank decided to give diplomacy a chance. He made peace with Francis and the society by apologising for Berkeley's outburst and by promising to support the society's efforts. But there was more to his strategy than just handing the palm to Francis Francis. On 1 January 1864, he wrote another, much longer letter to *The Times*. This one was headed 'Fish Culture in England' and was a masterpiece of doublespeak.

At 1,045 words, and with a 48-word postscript, it was somewhat longer than most page leads in modern broadsheet newspapers. Ostensibly it was written in reply to a piece on 'Fish Culture in France' which had appeared in the edition of 29 December, to correct the misapprehension that England still lagged behind other countries. In fact he was subtly reasserting himself as Britain's foremost authority on matters piscicultural. In his opening paragraph he made a list of English fish-hatching sites, placing Hampton first and Twickenham second. There were eighteen more (including one near Dublin), in all of which he asserted an interest. He followed this with a detailed, and not overly modest, account of his work with Stephen Ponder at Hampton. As a result of their enterprise in the previous year, he wrote:

> We let loose in the shallows in the neighbourhood of Hampton, Sunbury, Walton &c., 22,000 English trout, 6,000 Rhine salmon, 2,000 French trout, 3,000 ombre chevalier [charr], and 2,000 grayling, in all 35,000 fish, being when let loose about the size of minnows. Should there be any doubt as to whether these fish are doing well in the Thames, let me beg the reader to take a boat some warm morning next spring and paddle gently along the shallows above Hampton, and I shall be much mistaken if he does not see a considerable number of little silver-scaled beauties darting away like water swallows from the bank side.

The message to Francis Francis – 'put that in your pipe and smoke it' – was unuttered but clear. The trout eggs hatched at Hampton, Frank wrote, were all obtained 'by our own hands' from various parts of England, and he and Ponder had already begun to stock up for the new season. He invited 'all those who are interested in the beautiful process of hatching fish by artificial means' to visit Ponder's hatchery and see for themselves. The Acclimatisation Society, he then revealed, had also erected 'a large apparatus' at Twickenham. 'This apparatus is under the entire charge and management of Mr. Francis Francis.' (Code: 'Don't bother.') In case anyone had missed the point, he offered his own 'gratuitous services' – not only at Hampton but also at the zoo, the South Kensington Museum, the Crystal Palace and the *Field* office, 'for I am determined that the public shall know all that my experience enables me to teach them upon the matter'. The implied question – who needs Francis Francis? – had received its answer. Thus was revealed a new side to Frank's character. Amiable, kind and full of fun he might be, but people should think twice before crossing him. In his postscript he promised very soon to submit his observations on the breeding of oysters. He signed himself with full qualifications, MA, FRS,* &c., Late 2nd Life Guards, and gave his address as the Athenaeum Club, Pall Mall.

His relationship with the Acclimatisation Society otherwise continued amiably enough. Specimens received by the society's treasurer, J. Bush, and enthusiastically welcomed by Frank himself during 1863 included a pair of emus and a wombat.† But it was fish that would make the big news of 1864. A new character now enters the story. James Arndell Youl, an Australian-born agriculturist, had settled in England in 1854 and become a

---

* This must be either a mistake by the editor or an untypical conceit by Frank. It was a great disappointment to him that he was never made a Fellow of the Royal Society.

† Emu is now farmed and available online in America, for use in recipes including roast emu, oriental emu (with soy sauce, lime juice and ginger), emu kebabs and spicy roasted emu (with cardamom, juniper, coriander and garlic), emu chilli, emu meatloaf, chicken fried emu, emu tenderloin, emu spaghetti sauce and emu tacos. The meat is described as red, low fat, and perfect for absorbing marinades. Wombat is best served stewed.

leading light of the Australia Society.* Youl's greatest ambition was to introduce salmon and trout to his homeland. He had already made two attempts to transport live ova from England, in 1860 and 1862, but both had failed. The turning point came in 1864, and – who could be surprised? – at the hub of it was Frank Buckland. The official account of what happened was given by the British government's Inspector of Salmon Fisheries, William Joshua Ffennell, in his annual report. Ffennell, an Irishman and long-time campaigner for the improvement of salmon rivers, explained why Youl had failed. The eggs had been shipped in water troughs which, though cooled with ice, 'could not be kept in a state to enable the tiny creatures to live through the voyage, although they were hatched'. Then came the eureka moment. 'An experiment was then tried by embedding some ova in ice, at the Wenham Lake Company's Stores† in London, under the direction of Mr. Frank Buckland.' This had led to what Ffennell, restrained perhaps by the sober conventions of governmental language, described as 'a most valuable discovery'. It was found, he went on, that ova packed in ice could be held in suspended animation for more than a hundred days. When placed in running water, the fish then 'came into the world strong and healthy'. The opportunity was swiftly seized. A vessel was chartered, a plentiful supply of Wenham Lake ice laid in, and Australia got its salmon and trout. Ffennell in the end threw off all restraint:

Those who have carried out this very important undertaking so far, deserve the highest praise for their enterprising zeal,

* Re-formed in 1971 as the Britain-Australia Society.
† The Wenham Lake Ice Company imported and sold ice from Wenham Lake in Massachusetts.

perseverance, and scientific skill in this effort to introduce salmon into the waters of a far-off country, where they were before unknown; and with the name of Mr. Frank Buckland, before mentioned, who superintended the scientific department, the names of Mr. Youl and Mr. Wilson, of the Australian Acclimatisation Society, should be recorded as taking a foremost part, undaunted by every difficulty that arose, in the accomplishment of so interesting and important an object.

The inspector's enthusiasm was not shared by everyone. There were some for whom praise for Frank was bitter gall. Noses were out of joint. Jealousies flared. Francis Francis seethed like a cuckolded husband at the bedroom door. There would be consequences.

CHAPTER TEN

A Band of Players

Showmanship was central to Frank's character. From fairground hucksters to stars of the London stage, he empathised with all those like himself who felt driven to perform. High culture or low, it made no difference, and his respect for such people could lead to astonishing acts of generosity. A perfect example occurred while he was pike fishing with H. Cholmondeley Pennell on the Avon. The trip had begun with a dash by express train to Salisbury, followed by an ugly coaching accident which he was lucky to survive. But Frank thrived on drama. The memento mori did not keep him long from the riverbank, where he soon hooked and landed a powerful fish. Most fishermen in those circumstances would have rebaited and ridden their luck. But Frank was not most fishermen. He chose instead to lay down his rod, dissect the pike and think about how it might have experienced the world. Its predominant sense, he concluded, was sight. 'The fish I caught must have seen my bait at least ten or twelve yards off, for I saw him start from his lair in the weeds, and he came at it like a rocket . . . He could not have smelt it, though he has nostrils.'*

---

* This is particularly interesting when considered against his later thoughts about migrating salmon.

All this would have left most men at dusk with little on their minds but a hot meal, a glass of brandy and an early bed. Not Frank. While dressing for dinner at the Star Hotel in Fordingbridge, he was startled by a sudden, desperate cry for help. *Murder! To the rescue! Help, help!* He ran down the corridor, flung open a door and burst into a large room which the hotel's proprietor, a Mr Bill, hired out for sittings of the county court, lectures and, as on this occasion, theatrical performances. After dinner, Frank and Pennell hurried to join the audience (reserved seats sixpence, unreserved threepence), taking their seats during a love scene in which a lady was complaining about her gentleman's infidelities, which apparently had been revealed to her by her mother's overtalkative magpie. Frank was fascinated and appalled by the 'out-at-elbows' appearance of the production: the crude and flimsy sets (a living room, a forest and the actual room in which the performance took place) that had to serve for every play, and the way the actors coped in such straitened circumstances.

A piano, much out of tune, was played by a woman who once had been pretty, but whose face showed the rude lines of care and misery; though so poor and evidently an invalid, she was an excellent performer . . . The whole company consisted of six persons, and most extraordinary and clever shifts were adopted by them to carry out their acting. There was, however, such a peculiar careworn and poverty-struck looking appearance about all these poor creatures, that I was determined to learn more about them and their sad story.

It was a story fit for Dickens. The company's manager had once been a successful impresario but had lost his fortune by overreaching himself. His wife, the woman at the piano, had once been a star of the London theatre blessed with a magnificent singing voice. This had broken down one day, leaving her unable to speak above a whisper. She had borne ten children, of whom seven had survived, including a baby which lay in a corner while she played. Others in the company had endured similar downward trajectories, from early success to a perpetual struggle against homelessness and hunger. That very morning they had walked twelve miles with their props and scenery in a cart, then had spent more than they could afford on advertising themselves across the town. Their reward was an audience of fifteen, most of whom (in Frank's description) were 'the average specimens of thick-booted, grinning village boys, who were all clustered together . . . sitting upon the witness-box, used on county court occasions, like a lot of sparrows on a hedge top'. Bravely the actors struggled through a performance of *The Lottery Ticket*, then began a concert. When Frank arrived there were just two people in the reserved seats (actually a sofa and a few armchairs), and no one brave enough to applaud – an omission Frank immediately put right. He did more. While the performance was proceeding, he and Pennell went outside and bought tickets for everyone they could find – twenty-four in all – in the hope that the theatre might then look a bit less dispiriting. It was, nevertheless, a disaster for the company, the night's profit countable in pence. The actors tried to console themselves by believing the poor attendance was a seasonal blip: rural Hampshire was too occupied with the harvest to have time for the theatre. But it was cold comfort. The reality was that a family of five – husband, wife and three children – had ten shillings to keep themselves alive for a week, and single actors just three shillings and threepence. An actress told Frank she had not tasted meat for seventeen days and struggled even to afford bread and tea.

The Saturday audience was no better than Friday's. The house was nearly empty, yet the cast still played their hearts out, giving full value to the sparse few who had paid their threepence (the sixpennies were all empty). At the end, Frank and Cholmondeley Pennell invited the entire company to supper. This struggling band of players clearly affected Frank. In their doomed travails they perfectly exemplified hard-pressed and under-rewarded working people of every kind. It only strengthened his determination to improve the quality of their lives. Even more than the Acclimatisation Society, pisciculture gave him a chance to put food on the tables of the poor – a kind of democratisation unknown since cave-dwellers hunted as equals. For the rest of his life it would dominate his thinking.

His concerns were not all about diet. He worried about people's spiritual welfare, too, and about what they breathed as well as what they ate. More than thirty years before the founders of the National Trust fulfilled their promise to provide 'open air sitting rooms for the poor',* Frank believed urban people needed to see more of the countryside. Fresh air and a change of scene, he thought, were 'always grateful to the brain-worker, whether he lived a thousand years ago in Rome or this very week in London . . . Folks talk about hot and cold water baths, vapour baths, hot-air baths, Turkish baths, and other kinds of baths innumerable, but of all health-giving, invigorating baths, give me the good, pure, *fresh air* bath.' Frank himself liked to take the air in hard bursts, and nowhere better than on the footplate

---

* The words are Olivia Hill's. With her co-founders, Sir Robert Hunter and Canon Hardwicke Rawnsley, she established the National Trust in January 1895.

of the Great Western express to Didcot, on which he somehow wangled himself a ride. Ordinary passengers were advised to throw open the carriage windows 'and take your fresh-air bath as it pours thousands of cubic feet of oxygen and ozone into the "gas-pipe" of your lungs'. While Frank's published prose could sometimes slide into flippancy, his notebook was filled with sombre reflection.

I really cannot help thinking, that the Almighty God has given me great powers, both of thought and of expressing these thoughts. Thanks to Him, but I must cultivate my mind by diligent study, careful reflection in private, and intense and quick observation of facts out of doors, combined with quick appreciation of ideas of others. In fact, strive to become master mind, and thus able to influence others of weaker minds, whose shortcomings I must forgive.

I am like a ship at sea; my instructions are to do good and earn a livelihood. I must carry on board a cargo of information, with a sound ballast of truth, so that when a sudden breeze strikes the sail, and throws the ship on her side, the ballast will right her. There are so many waves – stormy, cold, white-faced waves of opposition – unkindness, and unconcern to encounter, that one need have good timbers.

This perfectly expressed what Bompas described as his 'simple-hearted, quaint, yet earnest' view of life and responsibility. 'Soapy Sam' Wilberforce, now Bishop of Oxford, thought the same. He found Frank 'very good fun . . . full of his natural history, and so simple and unaffected as to make him very charming indeed'. Well, yes . . . Unaffected Frank

most certainly was but the simplicity was illusory. The ability
to make complex ideas accessible is the hallmark of a good
writer, and Frank at his best was very good indeed. He thought
very hard about what he heard and saw, and never passed up an
opportunity to learn. One day, for no better reason than that
he had never been there before, he set off to visit Loughton,
near Epping Forest in Essex. He didn't much like the look
of it – 'a very sober, slow, stupid, cockney tea-and-hot-water
place it seemed to be' – but he trusted himself to find something of interest. His educators on this occasion were three men
whom he met carrying birdcages, and a boy bearing a stuffed
chaffinch. They were bird-catchers from Whitechapel, whose
secrets Frank swiftly unlocked with his 'usual key' of beer and
cigars. Songbirds in the nineteenth century were a highly prof-
itable commodity for the men who caught them, and Frank
wanted to learn how they did it. Nothing to it, said his new
friends. They would just stroll along until they heard a nightingale in a tree or a hedge, then they would trap it. First they
planted a stuffed bird – they called it a 'stale' – as a lure. This
would be wired to a piece of wood with a nail at one end which
they stuck into the bark of a tree. Around the stale went whale-
bone 'twigs' smeared with bird-lime, a sticky substance made
from boiled holly bark which the Whitechapel men valued at 'a
guinea a pound'. The finishing touch was a live 'call-bird' singing in a cage. The wild bird would think the singing was coming
from the stale. Down he would swoop to drive the intruder off,
and the bird-lime would do the rest. Unfortunately the trick
worked only too well, so an unnatural silence had fallen over
Loughton. 'The birds about here, sir, is pretty nigh all catched
up,' said Frank's informant. Frank noted the perverse effects of
supply and demand that would be so important in his thinking
about fish. It was a vicious cycle. Heavy demand shortened the

supply. Shortened supply heightened the demand. The rarer nightingales became, the more they would fetch and the harder the hunters would work to catch them.

On his way home Frank was saddened to see 'a foolish-looking cockney boy' with a nest of baby birds. All of them would be dead by morning. Pondering this, Frank made a direct appeal to his female readers: 'Ladies, pray forbid the idle boys from bird-nesting, and send your husbands and brothers after them *with the stick*; if you see them looking for nests, and you will be doing much good; for not only do the birds do the garden great benefit by eating insects innumerable, but consider how *you* would like *your* nest taken away.' Typical Frank: serious message, jocular style.

In common with his more thoughtful contemporaries (and many countrymen now), Frank was ambivalent about hunting and shooting. He enjoyed both, but felt people should spend more time watching and less with guns in their hands. 'We are too much in the habit of hunting, shooting, destroying, or otherwise tormenting the living representatives of the fauns and satyrs of our woods, lanes and hedges. How much greater would be the pleasure if we watched them a little more, and observed their instincts and their habits!'

In July 1864 the RSPCA invited him to judge a design competition for a more humane kind of vermin trap, one that would kill without torturing. Frank felt a particular concern for foxes, which he believed would gnaw off their own legs to escape from a trap. His evidence was gruesomely compelling.

It is not uncommon to hear of three-legged foxes being killed. The cause of loss of the third foot [*sic*] is generally a trap, and I am convinced that the story of foxes gnaw-ing off their foot is true. It may be urged that the pain of the self-amputation would prevent the fox from doing this; but it must be recollected that the trap, having cut off all circulation from the lower part of the foot, the latter would become dead, and numb to all feeling . . .

My friend Mr Bartlett . . . once had a fox consigned to him for stuffing which had only three legs . . . When he sent the animal home, the gentleman . . . was much surprised that the fox had four legs . . . Mr Bartlett [explained that] when he came to dissect the fox he examined the contents of his stomach, and found in it the missing foot, much gnawed by teeth, but still perfect enough to enable him to prepare it and restore it to its proper position. It was quite evident that the fox had bitten his foot off, and, in his agony, had swallowed it.

Frank despaired of taxidermists who failed to set up foxes correctly. 'In nine stuffed foxes' heads out of ten, the pupils of the eyes are made round like a dog's, and not elliptical like a cat's. The fact is, that a fox being a nocturnal animal, has a cat's eye, and not a dog's.' He observed nevertheless that a fox would wag its tail like a dog if it was pleased,* and he liked the fact that so many inns were named after what he now thought should replace the lion as England's national animal.

In the same year, 1864, Frank presented a paper to the Zoological Society on his experiments with cross-bred salmon and trout, wrote another paper identifying whitebait as (mainly) the young of herring and sprat, swapped notes on pisciculture

---

* His only evidence for this was a single incidence he had witnessed of a fox 'wagging his brush' at Edinburgh Zoological Gardens. He acknowledged that the fox was 'not often on terms of friendship enough with mankind to show his pleasure'.

with experts in Paris, and visited a salmon fishery in Galway where the owner, Thomas Ashworth, wanted him to explain the wholesale disappearance of newly hatched fry. Frank soon saw what had happened. The rearing pond was swarming with carnivorous water beetles. He did not just explain this to Thomas Ashworth, he *showed* him. Into a bottle of water went one beetle and two baby salmon. There was but a single survivor. This was typical of the Frank we know: plain-spoken, direct, unsentimental. But there was another side to him, less often revealed, which suggests the inner boy once in a while had to move aside for the inner poet. His first sight of mature salmon swimming in a Galway river roused him to an almost Wordsworthian lyricism.

Oh! You shining lovely creatures! At last, then, I see you free and at liberty in your native element. Mysterious water fairies, whence come ye? Whither are ye going? Why do ye hide your lustrous and beautiful figures in the unseen and unknown caverns of the deep blue sea? Why do ye shun the eye of mortal man? Hitherto I have seen only your lifeless, battered, and disfigured carcases mummied in ice and lying in marble state on fishmongers' slabs. Who could believe that in life you are so wondrously beautiful, so mysterious, so incomprehensible?

At the Galway fish ladder,* boy and poet became one. The boy wanted to know what it might feel like to be a migrating salmon. The poet wanted to describe it ('Oh! That I had scales and fins for five minutes, thought I . . .'). Frank stepped onto

---

* A fish ladder was, and is, a stepped structure enabling migrating fish to leap over obstructions such as weirs or dams.

the ladder, slowly lowered himself into the seething water and imagined himself to be a fish. '[I] congratulated myself on narrow escapes from the nets and the crevices below, and thought how very desirable it would be to get up to autumn quarters in Lough Corrib above.' But his inner salmon was going nowhere. He couldn't move. If he did, 'the water knocked me about like a wood-chip in a street-gutter after a thunderstorm'. All he could do was hold tight and hope a salmon would leap up the ladder past him. Instead he was assailed by a bailiff surprised to find this 'curious white-skinned creature' floundering half submerged on the salmon ladder. The bailiff – humorously, one must hope – offered Frank an alternative insight into the realities of life as a fish. 'Bedad, your honour . . . if I had got a gaff in my hand, I'd just strike it into your scales and see how you would like it.'

When regaled to readers of *The Field*, all this might have seemed innocent enough. For Francis Francis, however, it was a goad too far. He ridiculed Frank's findings in Galway, provoking a tart exchange of letters which culminated in a diatribe from Francis that blew away all pretence of civility. Enraged by what he saw as Frank's presumption, and angered by the Acclimatisation Society's suspension of his Twickenham fish hatchery, there was nothing left for him but to lash out. *What experience has Mr Buckland to offer that will bear comparison with mine?* In a letter to *The Field* he accused Frank of 'vague capriciousness', of doing more harm than good, of spoiling things for better men than himself, and of lacking professional etiquette.

[When] a fellow labourer is known to be engaged upon a subject . . . one ought not to descend upon it and take it out of his hands. But Mr Buckland's mode of procedure in *The*

Field has been to spread himself over every department, wandering hither and thither like a bee from flower to flower, or a Bedouin of the desert from richly-laden caravan to caravan. The moment a subject has been worked up by another person into notice, down pounces Mr Buckland, and like a blowfly, lays his eggs on the subject, and forthwith pervades it, often to the exclusion of the original proprietor.

There was little need for Frank to mount his high horse. Francis's abuse launched him into the saddle, willy-nilly. No gentleman, said Frank, could respond to a letter couched in such ungentlemanly terms, and no editor should have published it. His resignation from *The Field* was emphatic and irreversible. He had written presciently in his notebook: *There are so many waves – stormy, cold, white-faced waves of opposition – unkindness, and unconcern to encounter, that one need have good timbers.* His timbers having proved more than equal to the task, he was now ready to move on. In fact he had already done so.

His new passion was oyster breeding, but his modus operandi was the same as ever. First came observation. He studied the oyster beds and experimented with spatting (seeding the beds) and breeding. Then came education. He presented a report to the British Association, lectured at the Society of Arts, the London Institution and places as far apart as Clapham, Liverpool and Ireland. Less happily, he had a stab at commercial oyster farming. The vehicle for this was the Herne Bay, Hampton & Reculver Oyster Fishery Company, incorporated in July 1864 under Frank's own chairmanship. Fellow directors included Cholmondeley Pennell, a couple of knights (Sir Henry James Leeke, Sir Edwin Pearson), plus a local landowner, Reynolds Collard, and other men of substance.

The full story of this debacle is well told in an excellent little booklet, *Adventures in Oysterville: the Failed Oyster and Seaside Development of Hampton-on-Sea, Herne Bay*, written by Martin Easdown and published by Michael's Bookshop in Ramsgate (one of many books on local issues). The Herne Bay Oyster Company looked a certain winner. The nearby fishery at Whitstable was already hugely successful: Easdown records that in 1862–3 it sold 60 million oysters for a total of £91,000. Though the Whitstable company opposed it, Frank's reputation ensured the Herne Bay scheme got parliamentary approval. There were to be five breeding ponds, five oyster dredgers, a pier, storehouses, homes for the workers and a three-quarter-mile tramway connecting to the London railway. The company's territory extended to twenty-one square miles – seven miles along the coast from Swalecliffe to Reculver, and three miles out to sea. How could it lose? For a while it did seem to be fulfilling its promise. As well as farming its own stock, it imported oysters from France, the Netherlands, Portugal and Essex, and sent twice-weekly boatloads to Billingsgate. The pier was opened in 1866 by the Lord Mayor of London, Sir Benjamin Phillips, and celebratory glasses were raised at the newly opened Hampton Oyster Inn.* But the bubble was about to burst. The company had overreached itself and infrastructure costs left it badly underfunded. Frank scented danger and in 1868 he resigned. The company wobbled on for a few more years, paying next to no dividends while falling out with fishermen and squabbling with its more powerful neighbour. A succession of severe winters and a declining market did the rest. The Herne Bay, Hampton & Reculver Oyster Fishery Company

---

* Now the Hampton Inn. It is one of the last vestiges of the village of Hampton-on-Sea, which surrendered to the sea and was abandoned at the turn of the twentieth century.

went into liquidation and was finally wound up in 1884. All this is glossed over by Bompas in two brief sentences. Frank 'took part in the formation of a Company for the culture of oysters . . . from which he afterwards withdrew [and which] was not, for various reasons, successful'. Bompas was a loyal brother-in-law who had a blind eye for anything unpleasant. At least the oyster company, unlike poor illegitimate Physie, got a mention of sorts, but the curtain came down again over the nastiness with Francis Francis. On that, Bompas ventured not a word.

One can only imagine the rage with which Francis received the news in March 1865 that Frank had set up a fish hatchery for the Queen at Windsor and had been rewarded with the title Fish Culturist to Her Majesty. The endorsements piled up like Herne Bay flotsam. In April the Prince of Wales agreed to become the Acclimatisation Society's president, and Frank's old friend Bishop Wilberforce joined the council. Frank's, not Francis Francis's, was the voice everyone wanted to hear. He gave evidence to the Royal Commission on sea fisheries, and to parliamentary committees on the Thames fishery and the oyster beds of the River Roach. The Home Office listened to his advice on the effects of pollution on freshwater fisheries, and in May he agreed to become Scientific Referee for Fish Culture at the South Kensington Museum, where a couple of years earlier he had exhibited his fish hatchery. It would pay him fifty pounds a year. Thus would begin yet another chapter in this most overcrowded of lives.

While all this was going on Frank had been putting together the third volume (known as the Third Series) of *Curiosities of Natural History*, for publication in January 1866. Like its predecessors, this was an improbable compendium of stories, reflections and ruminations, the trivial and the profound all tumbling one over another. You could never know whether the next page would make you gasp, laugh or pause to think. What people *did* know was that no page could be skipped. He ranged from a Roman racecourse to a Yorkshire fishing match, church bells, fire precautions at the zoo, toothache in a hippopotamus, the pleasures of duck-hunting, the Uffington white horse (he calculated that the line from the lower jaw to the end of the tail was as long as the pavement in Trafalgar Square opposite the National Gallery), the pike's powers of sight and the whale's sense of smell. He explained how to open the jaws of an angry dog, catch a monkey, judge the weight of a tiger . . . There was even a bit of first aid. While trout fishing with Pennell, he had gashed a finger on a broken reed, causing profuse bleeding. He staunched the flow with the head of a bulrush ('in a few seconds the blood made a clot round the seed') and made a bandage with skin stripped from the back of a landed pike. Well, you would, wouldn't you?

Among those he visited in 1865 were the Dukes of Northumberland and Marlborough, the Lords Dorchester and Burleigh, and 'the Chinese giant', for whom Frank also held a dinner party (menu not recorded). Another visitor to Albany Street was the French giant Jean-Joseph Brice, nicknamed le Géant des Vosges, who advertised himself as Anak the King of Giants. Brice was rumoured to stand 245cm tall in his socks, though his actual height was a rather more modest 229.78cm, a fraction over 7ft 6in, and half an inch shorter than Charles Byrne. Not everyone thought Frank's fondness for freakery sat comfortably with his serious work on fisheries.

As it happened, the serious work had taken another turn. Frank had begun making plaster casts of fish. The idea apparently came to him during a visit to the Royal Society in Dublin, whose museum contained a number of extremely realistic casts. Frank straight away saw the potential. Though he was a great admirer of the taxidermist's art, he was also keenly aware of its limitations. Whereas a skilfully stuffed bird or animal could be made to look much like the living creature,* fish would lose their colour and look like their own ghosts. A carefully made cast, on the other hand, if accurately painted, would preserve all the physical detail of the fish *and* its living colours. Museum visitors would then have a very much better idea of exactly what it was that lurked beyond sight in lakes, rivers and oceans. They could see, too, how fish appeared at different stages in their growth cycle. His ambition, said Bompas, was 'to make the facts of pisciculture available to everyone'. In his usual fashion, Frank went at it wholeheartedly and learned from his mistakes. His journal entry for 28 November 1864: 'All day in the kitchen casting fish; the eel and the salmon; made some awful failures, but did not knock anything to pieces.'

His well-travelled fisheries display, including casts, now found a permanent home among the Animal Products and Food collections at the South Kensington Museum. Here was another good way to educate the public, and another good way to stir up trouble. The keeper of these collections, W. Matchwick, declared

Frank's casts of herring and baby sharks, now in the Scottish
Fisheries Museum

* The Natural History Museum in London still rather apologetically displays some of its early collection of stuffed specimens, but the best display I know about is in the Specola, the natural history museum in Florence.

in 1865 that the fish collection and hatchery were gratifyingly popular. But this wasn't enough to satisfy Frank. The museum's official report revealed that he had imported an entirely separate collection dedicated exclusively to 'the economy of fish culture and preservation'.

> It consists of hatching and rearing apparatus on a consid-
> erable scale, models of breeding ponds, weirs, fish ladders,
> apparatus for the transport of fertile fish ova to distant
> countries . . . diagrams and models of nets and apparatus
> used in the illegal destruction of salmon, illustrations
> of the natural enemies of salmon and trout, a series showing
> the growth of the salmon from the egg to the full-grown
> fish; a series of whitebait at various stages of growth, and
> of fish sold and eaten as whitebait, and an extensive series
> illustrating the growth and artificial cultivation of oysters
> from various parts of Europe and other countries; all of
> which is the private property of Mr Buckland.

Where in his hectic life had Frank found time to do all this? He certainly couldn't have done it entirely in the hours of daylight. One wonders at what time of night in Albany Street the lamps were dimmed and the candles snuffed, if indeed they ever were. But it might be thought – indeed, it *was* thought – that inserting a private collection into a public museum was a mite presumptuous. A representative of the museum's staff complained to the director that Frank had overstepped the mark. 'I think it inadvisable that an officer of the Department should be allowed to form and exhibit as his own property such a collection of objects as it is his duty to advise and assist the Department in collecting for itself, the prestige of the Department being used in procuring examples for what is

Frank in his Museum of Economic Fish Culture

in reality a private collection.' Everything in the display, he argued, should belong to the museum. This got short shrift from Frank, who pointed out that the exhibits were entirely his own, collected or built by himself in his own time and at his own expense. His meaning was plain. The museum should be grateful to him for having provided such a popular crowd-puller, not conspiring to rob him of his property. Both sides had a point (it is impossible to imagine a modern museum allowing its staff to own the exhibits), but it was Frank who won. We can't now know whether it was the strength of his argument or the power of his influential supporters that tipped the balance. Either way, the improbably named Museum of Economic Fish Culture was established at South Kensington under the curatorship and sole ownership of Francis Trevelyan Buckland. Bompas characteristically shrank from the unpleasantness like a snail with salt on its tail, gliding over it in one anodyne sentence. 'On receiving this appointment [Scientific Referee to the South Kensington Museum], he commenced at once transferring to the Museum his collection illustrative of fish and oyster culture.' Easy as that.

Frank's enthusiasm for casting famously would reach its apogee two years later with the Sturgeon in the Night – an episode so unlikely that it was still causing laughter in 1886, when the American magazine *Frank Leslie's Popular Monthly* remembered it with a cartoon. No cartoon, however, could match the hilarity of Frank's own account. 'I would advise my readers never to attempt to have a sturgeon stuffed,' he once wrote. 'Stuffed sturgeons are hideous monstrosities.' But he did acknowledge their versatility in the kitchen. 'It is said that a good cook can obtain beef or mutton, pork or poultry out of a sturgeon; in other words, fish, flesh, and fowl.' Nor was this the sturgeon's only virtue. If readers had a chance, he said, they should get hold of some of the large, knobbly bones that stud the fish's back.*

> They should be soaked for three or four days in water, and then boiled till all the flesh comes off. It will be found that the stud-like bones are most beautiful objects, being as hard or harder than ivory, with the outer surface indented and marked as though they had been carved by a Japanese artist . . . When set in silver, selected samples of these shackles form very beautiful ornaments for ladies' dresses . . . and I certainly would advise my lady readers who are always looking out for something new and pretty to try the effects of sturgeons' "shackles" when worn as ornaments.

---

* Frank's word for these was 'shackles', but they are more commonly known as scutes.

One Tuesday evening in April 1867, Frank received news
from a friendly fishmonger, Messrs Grove of Bond Street, that
they had a nine-foot, fifteen-stone sturgeon on their slab and
were willing to lend it to him for casting overnight. It would have
to be back in the shop by ten next morning. The first problem
was how to get the monster to Albany Street. Frank tried and
failed to manhandle it into a cab, then borrowed the fishmonger's cart. The next problem was how to get the fifteen-stone fish
out of the cart and into the house. Frank's first thought was to
shove it over the railings and down into the little front kitchen
where he did his casting. This was not a good idea: 'his back
was so slippery and his scales so sharp to the hands, that Master
Sturgeon beat us again'. Slapstick was a popular feature of the
mid-century Victorian music halls, but none can ever have imagined anything to beat this. *However*, Frank wrote,

> I was determined to get him down into the kitchen some-
> how; so, tying a rope to his tail, I let him slide down the
> stone stairs by his own weight. He started all right, but,
> 'getting way' on him, I could hold the rope no more, and
> away he went sliding headlong down the stairs, like an
> avalanche from Mont Blanc. At the bottom of the stairs
> is the kitchen door; the sturgeon came against it 'nose on'
> like an iron battering ram; he smashed the door open . . .
> and slid right into the kitchen, gliding easily along the
> oil-cloth till at last he brought himself to an anchor under
> the kitchen table. This sudden and unexpected appearance
> of the armour-clad sea monster . . . instantly created a
> sensational scene, and great and dire was the commo-
> tion. The cook screamed, the housemaid nearly fainted;
> the cat jumped on the dresser, upsetting the best crock-
> ery; the little dog Danny . . . made a precipitate retreat

under the copper and barked furiously; the monkeys went mad with fright and screamed 'Murder' in monkey language; the sedate parrot's nerves were terribly shaken, and it has never spoken a word since; and all this bother because a poor harmless dead sturgeon burst open the kitchen door, and took up his position under the kitchen table.

One of Frank's most valuable assets was his secretary, Mr. Searle. He it was who had the responsibility of actually casting the fish (hoisted onto the table with 'ropes and improvised mechanical contrivances') while his master kept a dinner engagement. With Frank out of the house, and the staff and menagerie calmed, the implacable Searle had the cast finished by two in the morning, and the sturgeon was back on Grove's slab by the

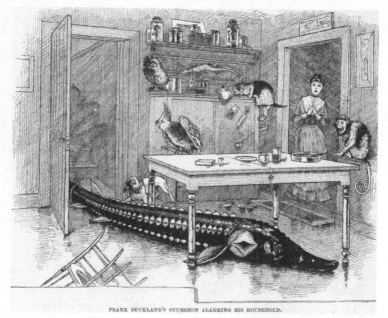

FRANK BUCKLAND'S STURGEON ALARMING HIS HOUSEHOLD.

Just another day at no. 37: a cartoon from *Frank Leslie's Popular Monthly* appointed hour of ten. The remaining unforeseen snag (which was evidently overcome, though Frank did not say how) was to extract from the kitchen a rigid plaster cast too long and too wide to pass through either the door or the heavily barred window. This disruption to the domestic affairs of no. 37 (as it had now been renumbered) would soon be eclipsed, however, if not actually forgotten.

A few days later the whale arrived.

CHAPTER ELEVEN

Mr Walpole's Pleasure

Students of the 1860s could be forgiven for believing the apocryphal Chinese curse, *May you live in interesting times*. 1866 is an arbitrary year chosen only because it is the point our narrative has reached, but it serves as well as any to give Frank's life a context. America was enjoying (or, in the southern states, enduring) its first year of peace after the Civil War. The infant superpower celebrated by adopting the Fourteenth Amendment, enshrining the equality of citizens under the law, a reform that went down better in the northern states than in the South, where white supremacists founded the Ku Klux Klan. In Sweden Alfred Nobel invented dynamite (patented in Britain, May 1867), and in America Andrew Rankin perfected the stand-up urinal. In Europe a fleeting Austro-Prussian War ended with defeat for Austria at the hands of Prussia and Italy. The Prussian Chancellor, Otto von Bismarck, survived an assassination attempt, as did (twice) Tsar Alexander II of Russia. Isambard Kingdom Brunel's *Great Eastern* succeeded at the second attempt in laying a transatlantic telegraph cable, 1,686 miles long, from Ireland's Valentia Island to Heart's Content in Newfoundland. The first message sent down the line was stiff but optimistic: 'A treaty of peace has been signed between Austria and Prussia.' Royal approval was conveyed from Osborne House on the Isle of Wight:

'The Queen congratulates the President on the successful
completion of an undertaking which she hopes may serve as
an additional bond of Union between the United States and
England.' In July, shortly after his eighteenth birthday, W. G.
Grace hit 224 not out for an All England XI against Surrey at
the Oval. Across the Thames in the same month, cholera took
yet another bite out of London's flank, killing thousands in the
East End. Other deaths that year included those of Thomas
Love Peacock and the one-time Surveyor General of India,
George Everest, for whom the mountain was named. Those born
in 1866 included Beatrix Potter, H. G. Wells, Herbert Austin
(founder of the Austin Motor Company), George Herbert,
fifth Earl of Carnavon (who would underwrite the search for
Tutankhamun), and the future Labour prime minister Ramsay
MacDonald. Disunity in the Liberal Party led to the resignation
of the prime minister, Lord John Russell, and a third prime-
ministerial term for the Conservative Earl of Derby (who would
die, worn out, only three years later, having uttered the bleakest
last words ever recorded: 'Bored to utter extinction'). July also
saw a royal wedding. Queen Victoria's third daughter, Helena,
married another future patron of Frank's, Prince Christian of
Schleswig-Holstein. Though the Wright brothers' first powered
flight was still thirty-seven years in the future, faith in heavier-
than-air flying machines was already sufficient to precipitate the
foundation of the Aeronautical Society of Great Britain (later
the Royal Aeronautical Society), which held its first public
meeting at the Society of Arts on 27 June. It was here that the
engineer Francis Herbert Wenham delivered the lecture 'Aerial
locomotion and the laws by which heavy bodies impelled through
air are sustained', which would inspire would-be aviators for the
rest of the century and beyond. Back on the ground, cutting-
edge technology still meant coal-fired steam power, which was

revolutionising marine transport without necessarily making it any safer. The risks began at the coal face – in December, explosions at the Oaks colliery near Barnsley, Yorkshire, killed 361 miners and 27 rescuers. There were two shipping disasters: the *Monarch of the Seas*, which sailed from Liverpool on 19 March with 698 passengers and crew and was never seen again; and, more famously (though it cost only 270 lives), the loss in the Bay of Biscay of the SS *London*, en route to Melbourne from Gravesend. This disaster is remembered chiefly because of another Victorian phenomenon, the indefatigable Scots poet William McGonagall. Nobody should believe his 'Tay Bridge Disaster' was the worst poem ever written until they have tried 'The Wreck of the Steamer "London" while on her way to Australia'. The first verse sets the scene:

'Twas in the year of 1866, and on a very beautiful day,
That eighty-two passengers, with spirits light and gay,
Left Gravesend harbour, and sailed gaily away
On board the steamship 'London',
Bound for the city of Melbourne,
Which unfortunately was her last run,
Because she was wrecked on the stormy main,
Which has caused many a heart to throb with pain,
Because they will ne'er look upon their lost ones again.

It grinds on for another eight steadily worsening, factually wrong but strangely endearing verses. What was remarkable about the Victorians was not just their appetite for innovation and improvement but their glad embrace of the eccentric and the weird. 'Characters' abounded, not the least among them Frank Buckland. What on earth would he come up with next? His unpredictability was part of what made him popular. No one ever knew what he might say or do, or what instruction he might give to his readers. McGonagall did not hit peak notoriety until the 1880s, so it is unlikely that Frank knew of him. This was a pity. McGonagall was a kind of literary freak (his deafness to criticism suggests Asperger's syndrome), and Frank collected freaks like other men collected stags' heads. Neighbours in Albany Street would crane their necks whenever a cab drew up. Who or what would step out next? Would it be eight feet or three feet tall? One head or two? Man, woman or beast? Across the threshold trooped much of the raw material for what would become Series Four of the *Curiosities of Natural History*. This would delight both his armchair admirers, who shared his love of outlandishness, and his critics, who liked having something to snipe at. What had dwarfs and giants to do with natural history? Frank's answer was always the same: if something existed, then it was interesting.

We may be glad that the exhibition of deformity in the twenty-first century is considered too offensive to contemplate. No one, not even the unfortunate woman herself, could now make a show of Julia Pastrana. It might be argued on the nineteenth century's behalf that the fear and cruelty of previous centuries had been supplanted by a simpler kind of fascination, lively rather than morbid. People didn't want to torture freaks or put them to death; they simply wanted to look at them. To Frank they were friends. He gave dinners in their honour, attended their weddings, celebrated the births of their children. His friendship with the genial French giant Jean-Joseph Brice began while he was still in the army, when Brice was a frequent guest at the barracks. 'It was a great fun', Frank wrote, 'to see our great, tall Life-Guard Troopers stand by his side, or walk under his arm, and look *up* to him.' His other special favourites were the 'Kentucky Giant', Captain Martin Van Buren Bates, and the 'Nova Scotia Giantess', Anna Swan. He first met them in 1871 when both were aged twenty-four. Captain Bates, he found, was a 'splendid-looking fellow, very unlike the pictures of the giant in the "fe fa fum; I smell the blood of an Englishman" legend'. But it was Anna who captivated him.

> I make bold to say that Miss Swan is the most agreeable, good-looking giantess I ever met; by her side I feel but a pigmy, for she towers far above my head; and an ordinary tall man, say a Life Guardsman, would look like a doll by her side. One never dares ask the age of ladies, nor their height either. Miss Swan is somewhere between seven and eight feet. I cannot say her exact height to an inch, but it is nearer eight than seven; at a guess (I hope Miss S. will forgive me) I should say seven feet six or seven inches is about the mark . . . Miss Swan is a native of Nova Scotia, is lady-like in manners and address, and would be a most agreeable neighbour at a dinner party . . . Captain Bates . . . is about as tall as Miss Swan, and a splendid couple they make when standing side by side . . . We hear rumours of the god Cupid having been seen.

The couple were married at St Martin-in-the-Fields in June 1871, and after a modest interval the new Mrs Bates would produce a baby whose birth weight topped 23lb. Her bridesmaids, Christine and Millie McCoy, conjoined twins better known as Christine-Millie or the Two-headed Nightingale, were close to normal height but no less extraordinary than the happy couple themselves, and they were just as much a part of Frank's colourful coterie of friends. The writer of *The World* magazine's 'Celebrities at Home' series was as impressed by

Huge love: Captain Martin Van Buren Bates and his bride Anna Swan,
who was a fraction under eight feet tall

the human element of the Albany Street ménage as he was by
the zoological.

It is [Frank's] delight to entertain celebrities on view in
the Town. This *penchant* makes him the idol of all the
children and stray waifs in the neighbourhood, who crowd
round the door when a party is expected, or clamber up
the railings to get a good view of the giant going in, or
the dwarf coming away. The due etiquette to be observed
at these feasts is at times perplexing. When Chinamen,
Aztecs, Esquimaux, or Zulus are the guests, the chief
difficulty is with the bill of fare; but the ceremonial
becomes complicated if Mrs. Buckland has to choose

which arm to take of the four owned by the Siamese
Twins; nor are matters put right by Mr. Buckland leading
the way with the Two-headed Nightingale; while much
discussion is needed to decide whether Mr. Buckland
should hand in Julia Pastrana (the hairy woman), or that
personage, by virtue of her beard, should take in the lady
of the house.

The 'Siamese Twins' referred to here were the originals, the
first ever to be so called, Chang and Eng Bunker. They had
been born near Bangkok in Siam – modern Thailand – in May
1811. Now in their fifties, they were veterans of the international circuit from which they had been irresistibly drawn
into the ever-widening Buckland circle. They were fully
formed, independently functioning individuals linked by a
cartilaginous bridge at the sternum and prevented from leading separate lives only by the fusing of their livers. If anything,
they were even more astute than the giants. Their career began
in 1824 when a Scottish merchant, Robert Hunter, saw them
swimming (or, by some accounts, rowing a boat) and did some
swift mental arithmetic. A deal with their parents followed,
and the teenage twins were soon embarked on a world tour.
Being far from stupid, they quickly realised they had no need
of a manager and could easily and more profitably run their
own business. At the expiry of Hunter's contract they seized
the initiative. In 1839 they bought a farm at Traphill in North
Carolina, where they became naturalised American slave
owners, adopted the surname Bunker and married two local
sisters, Adelaide and Sarah Ann Yates. The four-in-a-bed
marital arrangements were unorthodox but fruitful: Chang's
wife, Adelaide, produced ten children, and Eng's, Sarah Ann,
twelve. The American Civil War, in which the twins' sons

fought on the Confederate side, was unkind to them. Defeat forced them to return to the exhibition circuit and set themselves on the course that would lead, time and again, to Albany Street. But they were getting old now, and though Frank remained fond of them he was developing a stronger affection for Anna Swan's bridesmaids, Christine and Millie McCoy, the Two-headed Nightingale.

The Siamese Twins were certainly very wonderful people, but in Christine-Millie we have, I think, something more remarkable. The Siamese Twins are two old gentlemen somewhat advanced in years.* The 'Two-headed Nightingale' is composed of two charming young negress girls, who are united back to back by an indissoluble band. I do not recollect to have seen a more intelligent, ever-laughing happy face than that of Miss Christine. She has dark rolling eyes and jet-black hair, and though her features are those of the daughters of Ham, yet there is a quickness and intelligence about her that shows culture and education.

Millie is like her sister in face and in her charming manners. They live in perfect concord, and from long habit walk about and even dance, without any appearance of effort or constraint. They are called 'Two-headed Nightingale' because they both sing very well, and the duets they practise show they have good voices, which have been successfully cultivated. Their age is nineteen.

---

* In fact they did not have long to live. Chang began to drink heavily, and in 1870 suffered a stroke. He died in his sleep, aged sixty-two, on 17 January 1874, of a cerebral blood clot. The previously healthy Eng died three hours later.

The 'Two-headed Nightingale', Christine and Millie McCoy

Like the Siamese Twins, Christine-Millie had come to Europe from North Carolina, but there the similarity ended. The plantation owners Chang and Eng had possessed slaves. Christine and Millie *were* slaves, or at least were the daughters of slaves. Their parents, Jacob (an African) and Monemia (Native American), were owned by Jabez McKay, a blacksmith. When the twins were ten months old, McKay sold them to a showman for a thousand dollars. After being traded several times more, Christine and Millie found themselves in the ownership of Joseph Pearson Smith, who hired them out to showmen. They were still only three. Somehow during their travels Pearson Smith seems to have lost track of his investment, which, following an appearance at P. T. Barnum's American Museum in New York, was

soon on its way to England under the control of another couple
of chancers, 'Professor' W. J. L. Millar and William Thompson.
By late 1856, when the girls were five, Pearson Smith had picked
up their trail and now headed to London with their mother to
reclaim his property. Give him his due. As slave owners went,
he was not a bad one. Back in North Carolina the girls were
taught to read, write, recite, sing, dance and play the piano, skills
which would enable them to develop the stage act that made
them popular across two continents. In 1871 they gave a number of private performances to the most exalted of their many
admirers, Queen Victoria, at Buckingham Palace. The Queen
gave them each a diamond hairclip. The coincidence of Frank's
and Victoria's tastes is the clearest illustration of differently cali-
brated moral compasses, then and now. But who are we to say the
nineteenth century was wrong? Why should conjoined twins not
carn a living? And if they could sing and dance, then why should
that living not be earned on stage? There is no more effective a
stigmatiser of otherness than 'good taste', which was never a vice
of Frank's.*

The giants and the twins all worked at the upper end of
the show trade, capable of living up to their billing. The
giants were gigantic. The Two-headed Nightingale sang.
Lower down the scale, in darkened tents and the back rooms
of inns, the business was less scrupulous. But Frank loved
it just the same. He was a connoisseur of ingenious frauds
and could never resist a huckster's stall. He and Jean-Joseph

---

* After a spell with Barnum's travelling circus, Millie's poor health caused
the pair to retire some time in the late 1880s, when they returned to North
Carolina and busied themselves with charity work on behalf of African
American schools and churches. Millie died of tuberculosis on 8 October
1912. Christine followed her some hours later, helped on her way by morphine.

Brice enjoyed a good laugh over a fossil horse tooth which 'a gentleman' had sent to Frank as the fang of a giant, and he cherished an old story (dating from 1721) of a show-man exhibiting the bones from a porpoise's fin as those of a giant's hand. Frank's own forensic skills were not much tested by the 'Spotted Child', viewable for the price of sixpence at the Windsor Fair in 1861. The 'exceedingly pretty little flaxen-haired, blue-eyed, English girl' was indeed spotted as advertised, but only (as Frank's magnifying glass revealed) by reason of the strong solution of silver nitrate which had been sprinkled on her skin. Other stories were sadder. 'The Woolly Woman of Hayti' was advertised as a young beauty with long flowing hair but turned out to be a poor shrivelled old hag whose abnormal mass of hair was, like Julia Pastrana's, caused by a distressing disease. Or the all-too-well-named 'Australian Fat Boy', twenty-three stone and certain before long to give some heavy work to a gravedigger. But nothing amused Frank more than a good supply of 'mountebanks' at a racecourse or a fair. 'I never neglect any opportunity of learning how some of the more needy of the mixed multitude endeavour to gain a scanty living, and transfer a few coins from the pockets of their richer fellows to their own.' Some of the richest pickings were on Epsom Downs. On Derby Day one year he noted in quick succession a man 'with an enormous shock of wool-like hair . . . like a New Mexican savage' who, after years of practice and by pinching his nose, had taught himself to bray like a donkey; a pale man describing himself as the 'American diver', whose trick was to fish coins out of a water-filled tub with his lips; a grubby man who claimed to have been 'blown up by fire-damp' and who displayed a travel-stained model of a coal mine; a man with an electrical apparatus from which he dispensed shocks at a penny a time; a fire-eater; a stout acrobat self-styled as the 'Infant Hercules'; and a man who cracked stones with his fist.

None of this dented Frank's admiration for the more professional entertainments in London theatres. At the Alhambra in Leicester Square he was transfixed by the daredevil 'Omar', who walked upside down with his feet hooked into iron rings ninety feet above the stage – a performance 'really fearful to behold'. He applauded (and puzzled over) the skills of the escapologist 'Herr Tolmarque', whom Frank himself roped to a chair; and the 'Human Frog', who could smoke a pipe and drink a bottle of milk while submerged in a tank of water. Yet Frank never lost his taste for the penny-a-views, especially those lining the road outside the Islington Cattle Show. There he and Abraham Dee Bartlett paid their pennies to visit Fatima, 'the bodyless and legless girl', who was seen afterwards with a full complement of limbs skipping across the street to buy beer; the 'Irish Prize Wonder', a 'hideous fat woman who could hardly waddle'; and the 'Indescribable Female' and the 'Indiarubber Man', who turned out to be the dried body of a child about five years old, minus a leg and an arm, and a gymnast. Best of all was the 'Wild Man of the Woods', a 'very ugly nigger' who exhibited himself in a fried-fish shop. 'He gets his living by making hideous faces . . . The other accomplishments of the wild man are, I believe, that of biting off the heads of live rats and eating their bodies, dancing on red-hot irons, and drinking spoonfuls of lighted naphtha.'

Frank also felt great sympathy for the threadbare hordes of men, women and urchins who haunted the pavements peddling knick-knacks and novelties. A man in Leicester Square sold functioning microscopes (Frank calculated a magnifying power of twenty diameters), fashioned from pill boxes with lenses made by heating Canadian balsam. A glass-blower sold

glass pens, breast pins and peacocks; a metalworker offered a
five-part spit for roasting meat, so ingenious that if he had
a chance, said Frank, there was 'no telling how many benefits he
might confer upon mankind'. All these and more were offered
for a penny each, and Frank bought them by the handful. 'I am
forming a collection of various articles bought for one penny in
the London streets; and I would beg my reader not to pass by
these ingenious, poor, hard-working people, but to give them a
kind word of encouragement, and a little assistance by purchas-
ing a sample of their goods.'

Sometimes customers got nothing more for their money
than a good yarn. Abraham Dee Bartlett told Frank of a servant
girl who was sent to buy milk from a milkwoman near Albany
Street, and who was startled to find a stickleback swimming
in the jug. 'When the fish, all alive oh! in the cow's milk, was
shown to the old woman,' wrote Frank, 'she turned round to her
boy and boxed his ears. "Jimmy, oh! Jimmy, you lazy rascal,"
she said; "you never strained the water!"' If such petty frauds
were unexceptional, then so was the fish. Frank explained in
Natural History of British Fishes that sticklebacks made 'a very
great nuisance' of themselves by getting into reservoirs and
thence being pumped through the mains. He was, nevertheless, full of admiration for the tiny fish's skill in building nests,
works of art which 'exceed in beauty and complexity anything
that ever was thought of by the human mind'. In Prussia, he
reflected, sticklebacks were fed to ducks and pigs, so why not
in England? No opportunity should be wasted to reduce the
cost of meat.

The relationship of art with nature was another of Frank's
favourite tub-thumps. As with Landseer's lions, his taste was
for accuracy and he was contemptuous of anyone who imagined creativity could transcend reality. In October 1862 he

was astounded by the precision of a copper eagle made by a man named Phillips who exhibited it in Piccadilly. This was no mere 'reading-desk eagle' but an exact, life-size replica, every feather on its outstretched wings separately made and attached. 'I deeply pity from the bottom of my heart', said Frank, 'the poor "critic"' who could see nothing artistic in this. He proceeded to show the poor critic exactly how the job should be done. In May 1872 he paid a visit to the Royal Academy, where he intended to ignore the catalogue and 'see how far the Painter's art could convey what he really meant without the interposition of a printed description'. The pick of the exhibition, he reckoned, was a painting of an elderly grey-haired man with his hands tied behind his back, standing in front of three lions and four lionesses. He scanned the picture – side to side, top to bottom – like a sleuth at a crime scene:

The beasts are evidently very hungry, and they have slain and eaten a man not very long ago. There is a blood-stained spot on the ground, and I see a human femur (left side), a right tibia, a left humerus, and a bit of the pelvis, lying about. A bit of a scapula has flesh still upon it. All the bones are human. *Why* don't these savage and starved beasts instantly fly upon and kill this poor old man? Look at that three-year-old Lion coming round from behind the others; a sneaking, cat-like, but magnificent beast, worth £200 at least to our friend Jamrach, the animal-dealer. Look at that old Lioness snarling and showing her awful teeth and spine-covered tongue, and the old Lion licking his quivering be-whiskered lips. What is that curious light falling full and glorious upon the man? It is not natural, it is not the light of the sun or the moon, nor is it the electric light. By Jove!

I see. It is 'Daniel in the Lions' Den'. Splendid – grand. I congratulate the artist.*

Others did not get off so lightly. Frank vehemently despised a picture of 'a very thin, tall lady in the costume of Eve [i.e. naked], chained to a rock by her wrists' while 'an idiotic-looking young man' stabbed at a 'nondescript beast' apparently meant to be a dragon. Evidence of idiocy was everywhere. 'Fancy going out to fight a dragon with pigeons' wings tied to one's ankles . . . and with bare legs, like a Highlander!' This feeble hero was no butcher or anatomist either. His pathetic sword-thrust, a few inches into the dragon's mouth and just inside the ramus of the left lower jaw, would have done no more than annoy the monster and not hurt it a bit. 'It would simply transfix his parotid gland, if a dragon has a parotid gland . . .' A hunting scene ('not painted by a sportsman') was also swatted for its lack of realism. 'Who ever saw a hunted and beaten fox with clean fur like a lady's muff?' In another, an old man carrying five wolves' heads was insufficiently stained by his grizzly exertions. 'The hounds are much too clean: they don't look as if they have been fighting with wolves. If I had painted this picture I should have made the hounds with blood about their chaps, and one of them certainly going on three legs from a bite in the fore paw.' A picture of a North American Indian sitting by a fire on a prairie was simply ridiculous. The logs were too big, and where on a prairie would the man have found them anyway? Behind him in the distance his friends could be seen riding away, followed by his

* Frank does not identify the artist but it was almost certainly Briton Rivière, whose *Daniel in the Lions' Den* answers precisely to Frank's description and was painted in the same year, 1872. The painting is now in the Walker Art Gallery, Liverpool.

own loose horse. Frank supposed the man must be ill, for he showed no sign of injury. 'He has some internal disease, possibly peritonitis ... I suppose they [the other Indians] are going for the doctor.' In fact they were doing no such thing. Peeking at someone else's catalogue, Frank found the picture was called *Left to Die.** Failure glared at him from every wall: 'An animal, I suppose meant for a red deer, wounded – a bullet-wound on the left side. A wound at this part would not bleed much, because the scapula would act as a valve to keep the blood inside the thorax; and yet there is no end of blood. A miserable production, and as far as the animal goes not fit for a public-house sign. The rest of the scenery good.' After Frank's article appeared in *The Times* (where according to Bompas it caused 'much amusement'), he received a letter from Mr G. A. Sale, congratulating him on his 'very sensible and suggestive notes' on the pictures' zoological accuracy. 'These fifteen years,' Sale went on, 'I have been the art critic of the "Daily Telegraph", and am even now drudging at the canvases in Piccadilly; but I can assure you that your professed rough and ready critique has been to me a very valuable lesson, and I hope it may be one by which my colleagues in the ungentle art may profit.' Frank treasured this as 'one of the greatest compliments I ever received'.

The fourth volume of *Curiosities of Natural History*, published in 1872, would focus heavily on the giants, flea circuses, mountebanks and all the other human flotsam that bobbed across the surface of Frank's life. All this tended to undermine his reputation for seriousness, and yet the longest section of the book is the last – a lingering chronicle of his burgeoning love affair

---

* This can be identified as the work of the English artist Frances Anne Hopkins, painted in 1872 following travels with her husband in Canada.

with salmon. He had thrown himself at it like an impressionable schoolboy embracing a new and all-consuming passion – in other words, exactly like himself. *Everything* was a thrill to him. The love affair ('Mysterious water fairies, whence come ye?') had blossomed on the trip to Ireland that had so enraged Francis Francis. Though his accounts of it were often playful, his pursuit was intense, focused and single-minded. It was a quest for knowledge, no scrap too small to be noticed, but there was an art to it. Like a novelist he was always looking for the detail that would tell the story. He urged others to do likewise.

> Keep your eyes open, your intelligence sharpened; facts, facts, facts are what we want; for no one knows but that a fact, insignificant in itself – if it only be a fact – may lead to the most important results, not only in the cultivation of land, but also in the hitherto much-neglected cultivation of that which composes two-thirds of the earth, viz., the waters, whether inland or oceanic. He, therefore, who will discover any new fact relative to the natural history of useful fishes, as the salmon, trout, sole, turbot, and the bivalve puzzle, the oyster, will be conferring great benefit upon the public at large.

It could never be said that Frank failed to practise what he preached. He scooped up facts as a whale scoops up krill. Only by observation could the lives of salmon be unravelled and understood. This meant mapping entire river catchments, and it prompted him to return to a favourite theme – that people should take as much pleasure from watching wild creatures as from hunting them. He was appalled by the poachers, 'cowardly and unEnglish-like', who dragged clusters of hooks like grapnels through shoals of fish, trying to hook them in the side and maiming more than they caught. This happened most often beneath mill wheels, which themselves caused horrible injuries to the fish. It made Frank think about ways salmon might be prevented from getting into mill races: gratings, nets, underwater fencing, even *trompe l'œil* waterfalls. At least in Galway there were salmon for the poachers to hunt. Frank was even more distressed by what he saw, or failed to see, as he travelled around the country. Thousands of miles of upland streams were flowing deserted and salmonless. Why? Because they were blocked by mills and weirs, the deadening hand of industry snuffing out a species which, instead of gracing the nation's tables, was being treated as vermin. It was doubly senseless because it was unnecessary. All that the fish needed to negotiate such obstacles was a simple salmon ladder like the one he had climbed at Galway.

Frank listened carefully to what the Galway netsmen had to tell him. One of the great unknowns was how far salmon swam out to sea. The fishermen assured him they had seen them far out into the Atlantic, twenty miles or more, beyond the Isles of Aran. Other questions were easier. What was the secret of the salmon's sharp eyesight? Frank did exactly what we would expect him to do. He sliced open an eye, detached the lens and found it was better than his own magnifying glass for reading the small print of a newspaper. He also wanted to know how the fish kept themselves steady against powerful flows of water. Exactly how strong were they? He strapped a scale around his waist, harnessed a salmon to it and clocked an initial thrust of 23lb as the fish tried to swim away. It interested him that its efforts dropped off very rapidly: the second thrust was 20lb, the third 15lb, and then came no more serious thrusts at all. This was handy for fishermen to know. 'I am convinced', wrote Frank, 'that a salmon's escape from the angler's hook depends much upon the first plunge he makes, and that although his power to go against the stream be very great, yet he is very soon what is vulgarly called "done", if called upon to make extra exertion.' He made further tests of the fish's mysterious physiology. Crawling through a salmon trap, he sneaked up on a fish that had been hiding in the shadows. After a few failed attempts he managed to lay a finger directly over its heart.

> I could feel it distinctly beating and thumping through the skin . . . I then requested a friend standing upon the weir to take out his watch, and we thus ascertained that the pulse of this salmon beat 92 to the minute. I then tried [a second] fish, and found that his pulse was 103 to the minute. I also counted their respirations or the movements of the gills in breathing; the first fish respired 77 times in a minute, the second fish 79 times in a minute. I must, however, state that these fish had been running about the cruive [a weir or dam for catching salmon] . . . and I dare say they were in a bit of a fright, and their pulse beat quicker than usual, as I know from experience the pulse of a patient who comes to consult the doctor is often bounding away . . . from pure nervousness.

Another question: how long could fish survive out of water? Frank hung a freshly caught 10lb salmon in a landing net and was 'rather surprised' (it is unclear whether by the length or shortness of the elapsed time) that it ceased kicking in seven minutes and was dead in eleven. A fish knocked on the head by a 'priest' or killing-stick by contrast would die in twenty-five seconds. Much of Frank's time in Galway was spent flat on his belly, peering over the riverbank. He noticed the fish's habit of assembling in groups, and of reassembling in the same groups after they had been disturbed. '[Their] favourite position seems to be side by side, their fins almost touching, like cavalry horses in the stable at Aldershot; and when one of the party goes away he soon comes back and "falls" in with a regularity that would do credit to a soldier.' He noted how their habits changed at spawning time when, 'by a wonderful instinct', they spread themselves over hundreds of square miles, 'nature's object evidently being to scatter the supply of young fish over as large a tract of country as possible'. He noted, too, how the herding tendency reasserted itself when the smolts (young fish) set off downriver towards the sea. This sparked a lively correspondence on whether the fish travelled downstream head or tail first. One letter writer thought they went tail first so that they could take note of the landmarks – rocks and tree stumps, perhaps – to help guide them back from the sea. Frank himself thought they went tail first over waterfalls to protect their heads. Another correspondent, J. H. Nankwell, believed the fish kept their heads out of the current to save themselves from drowning.* What mattered to Frank was observable fact – the *what* rather than the *why* or the *how*. It seemed to him that salmon, like birds, had mysterious powers of navigation, and a mysterious ability – even at high speed through the most intricate labyrinths of weeds, rocks and roots – to avoid collisions. The *why* and the *how* were unknowable to any but God. Another point of disagreement was how long smolts spent at sea. Experiments with marked fish at Galway revealed no fixed

* Nankwell wrote: 'In the fish, as soon as the mouth is opened, the water rushes in to fill the vacuum so formed, and then, the mouth being closed, the muscles (Pharyngeal?) contract, and send the fluid out over the gills. The water thus falls in with the general current and is carried off; on the other hand, if the animal has to pump the water back against a strong current, the muscular effort to do so must be increased manifold (?), and the creature must feel more or less of what we call dyspnoea [breathlessness].' Frank's basic observation was correct. Smolts are carried seaward, tail first on the current – maximum mileage for minimum effort.

pattern. Some, but not all, stayed out for a single year before returning as grilse. Others lingered two years or more. This very randomness, Frank argued, was fitted to nature's purpose.

It appears to me to be a law of nature that the salmon (say in an individual river) shall never be *all* subject to the same influences at the same time, and this is as a protection against their numerous animate enemies, as well as pollutions . . . Nature seems to say, 'I will send some of you youngsters up the river in 1856 and some of you shall stay in the sea till 1866; so that if the first lot of you get destroyed, there will still be a second batch on hand to take your places and keep up the supply in the river for future years.' Again, in our own species, we do not all take our stand in the battle of life at the same age. Some boys are sent to 'cut their own grass' at eighteen, some not till twenty-two or twenty-three. Young ladies, as well, do not always 'come out' into society at exactly the same age.

But nature could not do its work without the active cooperation of humans. Waters had to be stocked and seeded, otherwise the 'water farmer' could look forward to barren harvests. The fish needed to be protected from poachers at spawning time. They needed clean water, and they needed help to overcome man-made obstacles. Easy to say, and in some places perhaps not too hard to achieve. But in the rivers of the muck-and-brass industrial heartlands, foaming with chemical effluents and sewage, clogged with locks, weirs and other barricades haunted by poachers, salmon might seem about as likely as migrating elk. A Royal Commission in 1860 had painted a picture of unrelieved bleakness. '[The] considerable diminution of salmon in the rivers of England and Wales was fully substantiated. In some rivers the fact was patent and notorious. Salmon formerly abounded, but had almost or altogether ceased to exist . . . Weirs, fixed nets and fish-traps, insufficient close time, pollutions, destruction of unseasonable or of immature fish, the want of an organised system of protection, and confusion and uncertainty of the law, were the chief causes.' This was followed in 1861 by a Salmon Fisheries Act which was supposed to replace the old mish-mash of confusing and often contradictory bit-and-piece legislation, some of which dated back to the Middle Ages. It specified annual close seasons, made fish passes compulsory and harrumphed about pollution. The problem was that at local level it made no provision for anyone actually to enforce it. At national level, responsibility fell upon two newly appointed Inspectors of Fisheries answerable to the Home Office, Frederick Eden and William Joshua Ffennell.

It was against this endless churn of disaster, degradation and dwindling optimism that Frank went on rescarching, reporting and proselytising. Since his falling-out with *The Field*, he needed a new platform from which to report his findings and air his views. He had wasted no time in fixing one. Supported by the publishers Chapman and Hall, and backed by his friends Higford Burr of Aldermaston Park and William Joshua Ffennell, he now had a whole new magazine of his own. The first issue of *Land and Water*, launched as a competitor to *The Field*, was published on 27 January 1866, price sixpence. Frank's presence as a principal contributor and editor of the natural history pages guaranteed its popularity. His first words to his readers were the enfolding embrace of a shepherd to his flock:

Let none think himself unable to advance the great cause of practical natural history. Thousands of Englishmen and Englishwomen have knowledge and experience, acquired by their own actual observation of useful facts related to animated beings, be they beasts, birds, insects, reptiles, fishes or plants. Friendly controversy and argument is invited on all questions of practical natural history, and although the Odium Salmonicum not unfrequently assumes more virulence than even the Odium Theologicum* of the good old days of faggot and stake, no writer need fear that his pet theory shall be ruthlessly set on fire, or that his arguments shall be decapitated, without fair and patient hearing.

In this he was true to his word. Readers were only too eager to send in their thoughts and observations, and Frank's lively exchanges with them were essential to the magazine's appeal. 'It would be hard to find more entertaining reading', wrote John Upton in *Three Great Naturalists*, 'than his answers to correspondents in the early days of the paper . . . He considered no trouble too great if he could impart knowledge to the public . . . When some snails were sent for his inspection, he carefully fattened them on lettuce, cooked them and ate them. He found them excellent – an opinion which the present writer can unreservedly endorse.'

Influential though Frank was, however, he had only the status of *enthusiast*, a brilliant cheerleader but powerless to be the all-conquering champion that England's rivers so urgently needed. He could not be satisfied with mere advocacy: he wanted to be the architect of events, to turn thoughts into actions. There was prophetic irony in the way the opportunity arose. Inspecting salmon rivers was physically demanding – too much so for one of the two

---

* *Odium theologicum* means literally 'theological hatred', or hatred caused by religious disputation. The joke *Odium salmonicum* was Frank's thinly veiled reference to the enmity of Francis Francis.

Home Office inspectors, Frederick Eden, who became ill and could not continue. Frank got wind of this from an old Oxford friend, the Financial Secretary to the Treasury, George Ward Hunt (a future Chancellor of the Exchequer), whose letter to Frank, dated 11 October 1866, survives in the scrapbook: 'Dear Frank, I hear there is just a possibility of a vacancy in the Fishery Inspectorship – I advise you to write a line to Mr Walpole* at the Home Office who has the patronage, asking him to consider your claims – I am going to him this morning to urge them but I recommend a formal application.' Frank did not need to be asked twice, and Hunt was as good as his word. Frank's diary entry for Wednesday, 6 February 1867 described one of the greatest days in his life:

This day I was appointed Inspector of Fisheries. I had been invited to dine at the Piscatorial Society in St James's Hall, and was sitting on the left hand of the chairman (Mr Sachs), when John brought me in a letter as follows: 'Home Office, February 6, 1867. Sir, – Mr Walpole has desired me to inform you that he has much pleasure in appointing you Inspector of Salmon Fisheries in accordance with your wishes. I am etc, S. Walpole'.† – When I read this I felt a most peculiar feeling, not joy, not grief, but a pleasurable stunning sensation, if there can be such a thing. The first thing I did was to utter a prayer of thanksgiving to Him who really appointed me, and who has thus placed me in a position to look after, and care for, His wonderful works. May He give me strength to do my duty in my new calling! I said not a word to anybody, but in a few minutes I had to make a speech, to propose

---

* Spencer Horatio Walpole, the Home Secretary.
† The Home Secretary's elder son, also called Spencer Walpole, who served as his private secretary.

the health of the prize-givers. I alluded first to the cultivation of the waters, and then to my excellent father's endeavours to do good, saying it was my wish to honour his name, and do my own duty in my generation. I then read out the letter, which was received with great applause. Thus, then, I have gained the object of my life. Surely fortune favours me with great luck; and I am very thankful for it. When I got home I found the house in a state of uproar; all the servants, the monkeys, Danny the little dog, the parrot, and the cat, with paper favours on; M. and L. were also here with favours on, and all much delighted by my appointment.

CHAPTER TWELVE

Torrents of Filth

F rank was used to driving himself hard. Curating a museum, editing and writing *Land and Water*, collecting ova, ministering to casualties from the zoo . . . keeping all this going was a fourteen-hours-a-day calling, exhausting even for him. In the spring of 1866 he took a few days' rest at Herne Bay, where he wrote to his sister: 'I have been working double tides lately and am now doing the work of two people . . . In both cases whip and spear and no assistance.'

On top of all this he now had to bear the responsibility, and the colossal physical challenge, of his official duties. On his first day at the office, 12 February 1867, the new Inspector of Salmon Fisheries wrote so many letters that he ran out of paper and 'knocked the bottom out of the inkstand'. But that was just the start. At home afterwards: 'Made cast of the big 38lb fish sent me by Quelch, who came in to see him cast; began at 9, finished at 1.' Presumably then he permitted himself a few hours' sleep. Next day he was back at the office, where he stayed until four in the afternoon. On the 14th it was a *Land and Water* dinner. 'All went off well; thankful I was in excellent humour for it, surrounded by many excellent and kind friends . . . Set up paper, then to the Office; gave an opinion about Fishery districts; came back by 4, and wrote some forty letters.' Four days later he travelled west to begin his first official inspection of a salmon river, the Exe. It was a perfect introduction to a labour that would consume the rest of his life. The inspection took seven days, but it would take years of effort before the river returned to anything like normal life. During the reign of Elizabeth I it had been, in Bompas's magnificent phrase, *fluvius piscossisimus*, but now it was an all but salmonless waste. Frank spent his seven days prowling around the eight weirs that had throttled the fishery, noting the spots where the bewildered fish gathered beneath them, watching them struggle to find a way upstream and deciding how best to help them.

The Exe was by no means the worst challenge he had to face. First he had to cope with the loss of his friend and fellow inspector William Ffennell, who died on 12 March and was replaced by Spencer Walpole. Then he had to identify his priorities. These were easy enough to state. He spelled them out in his first annual report, published in July 1867. Salmon, he said, needed unimpeded access through clean water to suitable breeding grounds and a safe place to feed. Simple as that. 'Were all these conditions combined, without let or hindrance, and the enemies of the salmon at the same time kept in check, there might hardly be any limit to the number of fish available for human food.' All they needed, in other words, was a miracle. It reminds me of a drawing by the brilliant young cartoonist Timothy Birdsall, who died aged twenty-six in 1963. It shows two industrialists gazing at a smoke-blackened sky. The older man is addressing his evidently more idealistic but unworldly companion: 'Smokeless zone? Nay, lad, tha can't tamper wi' nature.' In the industrial heat of the nineteenth century, factory owners took a similar view of the rivers. Pollution-free water? Nay, lad. Forget it.

In 1867 Frank reported that only five salmon rivers – the Otter, the Ribble, the Lune, the Glaslyn and the Conway – were acceptably clean. For the rest he could report only torrents of filth. Five were polluted by gasworks, six by paper mills, six by lead, five by mining, three by coal dust, three by sewage, three by tanners, three by 'carpets, chemicals, &c.', and two each by the wool and dye-making industries. 'Manufacturers of all kinds of materials,' he wrote, 'from paper down to stockings, seem to think that rivers are convenient channels kindly given them by nature to carry away at little or no cost the refuse of their works.' It was not just fish that suffered. Eighteen years had passed since William Buckland preached his firebrand sermon in Westminster Abbey, excoriating the diehards who refused to see the connection between foul water and fatal disease. Frank employed the same argument. 'Impure and polluted water', he advised, 'will encourage disease, especially cholera; pure water will disarm disease of its power, and at the same time be available for growing excellent human food.'

But even that was not the worst of his problems. Weirs were the perfect piscatorial prophylactic. If the fish couldn't reach their spawning grounds, then there would be a birth rate of zero. The obstructions, Frank reported, were 'formidable in number, and formidable in construction'. On the Severn and its tributaries there were seventy-three weirs; on the Taw and Torridge seventy ('forty miles of beautiful spawning ground blocked out!'); on the Wharfe fifteen; on the Derwent eight; on the Dee five, with more on the tributaries; on the Nidd eight, Swale six, Ure four, Ouse two . . . Very few had none at all. With the patience of a schoolmaster stating the obvious, Frank set out the arguments: 'A river serves two purposes; it may, firstly, be made to work milling power; and secondly, it may be made to produce fish.' But there was no need for conflict; no reason why conservators and millers should not succeed in both their enterprises, 'particularly if a friendly spirit be brought to bear.' Friendly spirit was Frank's speciality. His engaging personality had propelled him through school, through university, through medical school, the army, London society, authorship and now into the foothills of government. The Buckland wit and charm, more even than the law, were the most powerful weapons in the salmon's protective armoury. Frank explained how fish could be helped over a weir at no inconvenience to the miller. 'A sloping trough with frequent bars across gave the principle of a water-ladder, moderating the flow of water, and forming a series of pools from one to another of which the salmon could swim or leap, and so pass over the weir.' This ladder should have a slope of six or eight feet for every perpendicular foot of fall, and should begin from the point beneath the weir where the fish would naturally congregate. Problem solved! The miller could have full use of the water during the day, and the fish could have it at night and all day on Sundays. It should not be a case of bread *versus* salmon, but of bread *and* salmon. Who could argue with that?

As the years rolled by he would cross and recross the country, making his pitch over and over again, letting no stream lie forgotten. Much to his regret he had no statutory power to modify weirs, but he had a genius for persuasion. Bompas put it perfectly: 'He seemed to have power to charm the reluctant; many who came to listen, suspicious, dogged and perverse, soon yielded to his genial tact and wit, and became themselves inflamed with piscicultural ardour.' When millers understood that salmon wouldn't hit their pockets, fish ladders at last began to appear. Progress was not fast. It would not be until 1875 that a significant catch would be taken from the Exe. But progress it certainly was, and it was Frank who made it happen. His interventions were often straightforwardly practical. When a fish ladder on the Stour failed to work he 'did a Galway' and climbed down to get a salmon's-eye view of it. 'His sympathy with the salmon soon discovered and cured the defect,' said Bompas, 'and then, at the cost of £15, he placed a cheap and simple ladder over the upper weir, which the fish fearlessly swam up, as if they knew it was planned by their best friend.' Every little helped.

Another essential was protecting the fish at spawning time. All the good work would be wasted if they died before they had time to breed. In all Frank's writing I have found not one word of sympathy for poachers, though I dare to hope he would have forgiven hungry villagers for taking the odd rabbit, and would have deplored the magistrates who imprisoned them or sent them to Tasmania. For those who waited with spear and trident at the spawning grounds, he had nothing but loathing. *Notes and Jottings from Animal Life* contained this account of a visit to the River Caldew, a tributary that joins the River Eden near Carlisle:

Looking over the river bank towards the Factory pool, what should I see but an old woman standing on the bank just in the very middle of this happy spawning-ground. The old woman was picking stones out of the river bed with a sort of curved pitchfork; she had a huge petticoat, under which probably pockets existed, so that without doubt she often picked out other things from the river besides stones. I am sure that old woman was a salmon poacher with a stick in her pocket, and tapes round her waist on which to hang under her petticoats the salmon she gaffed. Stone picking of course had driven away any fish that might have been on these beds. A curious-looking party was this old woman, and when I came home I found her portrait in my little marmoset, only the little monkey is much prettier than this horrible old Cumberland native.

Even when spawning was successfully achieved, Frank said, the odds were against the fish. 'The trout then comes to eat the eggs; next a whole swarm of flies and insects; then the water-ouzel [dipper], who goes to eat the flies, is shot by ourselves, under the idea that the bird is after the eggs, and not after the flies. Other enemies come; the jack [young pike], the otter, who follows the little salmon on their way to the sea, where the angler fish lies in wait for them. The result is that not one egg out of ten thousand ever becomes food for man.' Frank deplored the wastefulness of anglers who took spawned fish, or kelts, which in their emaciated form were not fit to eat. Far better to let the exhausted hens return to the sea and fatten up again.* To prove his point he collected two specimens, a kelt and a fresh-run fish each thirty-nine inches long, and put them on the scales. The kelt weighed 10lb 2oz, the fresh-run fish 20lb. Frank made plaster casts of both, and put them in his museum.

Before the next annual report in 1868 he had visited another twenty-six rivers, affording each one the same weir-by-weir attention he had given the Exe. The report complained of sewage in the Stour and in the Thames, where he had himself suffered bitter disappointment. Despite the thousands of fry he had released at Hampton, only one or two fish had ever been caught. The wider picture was no better. On seventeen supposed salmon rivers, fish were absent from 11,600 square

---

* It is now thought that only around 5 per cent of kelts that return to the sea survive to spawn again. Most male fish die after spawning.

miles of water. Figures from Billingsgate market told the same story. Yearly sales of Scottish salmon varied between 750 and 2,100 tons, with an average of 1,100, and Irish salmon averaged around 350. The combined sales of English and Welsh salmon had gone up too, but only from 55 tons a year to 88. Even so, it meant that annual sales in England and Wales were worth a far from insignificant £100,000 (£9.5 million at today's values). 'To accomplish these results', wrote Bompas, 'required ceaseless personal exertion. To stir up the inert, to combat stolid prejudice, to harmonise conflicting rights, to infect others with a spark of his own enthusiasm; this was Frank Buckland's daily task.' A vital adjunct of this task was Frank's writing. In the preface to *Log Book of a Fisherman and Zoologist*, published in 1875, he explained how he found the time.

I write most of my articles (especially the long ones) in railway trains. My Official duties as Inspector of Salmon fisheries . . . necessitate frequent long journeys . . . During the otherwise tedious railway journeys I always do a great deal of writing. Going north I post my MS. – begun at Euston or King's Cross and written during the journey – at York, Newcastle, or Carlisle; and when going in other directions I post at Salisbury, Exeter, Gloucester, Worcester, Shrewsbury, Canterbury, or Chester.

Thus, a railway carriage to me becomes a most agreeable studio where I can write without fear of being called upon to attend to other matters, while the cold-air bath, as the Express rushes along, invigorates the memory and renovates the mental powers. The articles which my friends tell me are my best are generally written in the railway train.

This may be true: one cannot tell. Burgess accused Frank of 'triviality', which I think is a bit unkind, but the sheer volume of his output did mean occasional lapses in form. Neither was he free from error. But he was a pioneer treading new ground in a largely unexplored backwater of natural science, and he respected others' opinions. The same can be said for only a few men of his or any other time, and few remain such a pleasure to read so long after their deaths.

The other great generator of popular enthusiasm was his museum, which mopped up most of whatever time he had left. He was not shy of mentioning this in the inspectors' official report: 'Being exceedingly anxious that the public should have a good idea of the nature and importance of the British Salmon Fisheries, I am happy to be able to report, that I have now got together a Museum of Economic Fish Culture; and my collection is open free to the public at South Kensington.'

It was not just the public who came. Queen Victoria herself toured the museum in May 1868 with her son Prince Arthur and son-in-law Prince Christian. Frank seems to have struck up a close relationship with Christian, who sought his advice on the management of deer in Windsor Great Park, and to whom he would dedicate *Log Book of a Fisherman and Zoologist*. There was much for the royal visitors to admire. South Kensington was the invariable destination for anything Frank found particularly interesting or unusual. The famous sturgeon was already there. New arrivals over the years would include the head of the turtle on whose back young Frank had ridden in Christ Church pond; the preserved gullet of an otter; coloured wooden replicas of a 6ft 3in halibut and a monster basking shark, 28ft 10in long, which had been washed up near Ventnor; the cast of a small tuna (the fish itself subsequently eaten); a stuffed sea eagle; a stuffed seal cub; the stuffed head of a minke whale. All these might be said to have some relationship to the marine environment. But there were many others whose only link to 'economic fish culture' was Frank himself. There was the spoof 'King Charles's parrot'; a sword dredged from the Serpentine; a man trap; the skin of a boa constrictor. More outlandish items included a stuffed seagull with an Act of Parliament* in its beak, and a mouse with its head trapped inside an oyster. Frank had a theory about this. He identified the oyster as a 'well-grown five-year-old Whitstable native' and decided it must have been 'put on the kitchen or larder floor; he then opened his shells as oysters always will do under such circumstances; and the mouse – a young and inexperienced mouse – put his head in between the shells to nibble the beard of the oyster, who instantly closed them and made the mouse a prisoner'. The mouse was rivalled for bizarreness by a rat which had been given to Frank by an official at Waterloo Station while still alive. This unlucky animal had somehow got a section of a pig's thigh bone – 'a little larger than a gentleman's full-size finger-ring' – stuck like a collar around its neck. Frank puzzled over this: 'The only way I can account for the bone being on the rat's neck – unless it was put there by human hands – is that when he was young he had been stealing and gnawing at a rasher of ham. During his work he had, unfortunately for himself, thrust his head through the ring-shaped bone, the set of his head and the size of his ears not enabling him to pull it off again.' As the rat grew, so the ring tightened and caused its neck to swell. Frank did his best for the animal. He sponged the inflammation with warm water but had to give up any hope of keeping it alive.

* Probably the Seabirds Preservation Act of 1869, the first Act of Parliament to protect wild birds in Britain.

A quick mercy killing followed, and the rat was preserved for the museum in a bath of spirits.

Frank hated to see any animal suffer. In 1871 he was enraged by the sad condition of pigs at Newcastle market, so thirsty that they 'sucked and fought for every drop of dirty water they found in the gutters or puddles'. An indignant article followed: 'Upon all the faces of these worn-out pigs there was but one expression, meaning, as plain as a pig's face can speak, *water for mercy's sake, give us water!*' He appealed to the city authorities and wrote to the secretary of the RSPCA, with the result that Newcastle market was swiftly supplied with troughs. On top of all else, Frank was one of the earliest practitioners of campaigning journalism. But he was not one of those animal enthusiasts whose sympathy was reserved for other species. From both his parents he had inherited a sense of duty towards his fellow humans, reinforced by his Christian faith. His colleague Spencer Walpole described another example of his kindness: 'One night . . . he found a poor servant-girl crying in the street. She had been turned out of her place that morning, as unequal to her duties; she had no money and no friends nearer than Taunton, where her parents lived. Mr. Buckland took her to an eating house, gave her dinner, drove her to Paddington, paid for her ticket, and left her in charge of the guard on the train. His nature was so simple and generous that he did not even seem to realise that he had done an exceptionally kind action.' No doubt after seeing the poor girl safely on her way, Frank returned to Albany Street and cast another specimen. One of the most spectacular of these arrived at Regent's Park on Saturday, 27 April 1867, when it was tipped out of a van onto the dissecting-room floor. 'It was a fine whale,' wrote Frank, 'but not so large as I should have wished.' The whale was one of several which had been caught in the Firth of Forth. Though small for a whale – 11 ft 6in long and 7 ft 6in in girth – it was still quite big enough to set Frank a serious test. Me too, for that matter. One of the difficulties of watching Frank from a distance of 150 years is that nineteenth-century nomenclature was very different from the names in use today. For this reason it is not always easy to know precisely what species Frank was talking about. Patrick O'Brian's wonderfully drawn character Stephen Maturin, whose encyclopedic knowledge is one of the great joys of the Aubrey–Maturin sea novels, presents the same challenge. O'Brian's dedication to authenticity in everything from a ship's rigging to the language of the sailors precludes him from offering his readers any help with translation. It doesn't matter. These are novels, and O'Brian wants us to use our imaginations. With Frank it is rather more important to know what he meant. He described his whale thus: 'There can be no doubt, I think, that my specimen is *Globicephalus deductor* or the ca'ing whale. His head resembles somewhat a huge round glass carboy as used by wholesale chemists, or the forehead of a person afflicted with hydrocephalus.' This sends me on a merry dance. In 1865 the name ca'ing, or caa'ing, whale belonged to a species then known as *Phocaena melas*, also known as the 'round-headed porpoise'. The *Daily News* of 23 August 1879 reported more than a hundred ca'ing whales driven ashore in Shetland, but the Latin binomial this time was *Delphinus deductor*. I discover as I go that ca'ing/caa'ing is a corruption of 'calling', meaning in this context 'driving', but this is interesting rather than useful. *Globicephalus deductor*, *Phocaena melas* and *Delphinus deductor* are all unknown to the International Union for the Conservation of Nature and to the Catalogue of Life, the online database of all known species. 'Ca'ing whale', however, does throw up a clue. Among the synonyms is 'pilot whale'. And here I believe we have the answer. By sifting the taxonomic references and comparing

Frank's description of the glass carboy or hydrocephalic head, along with his observations about the creature's eyes and fins, I am as sure as I can be that his ca'ing whale was the animal now known as the long-finned pilot whale, *Globicephalus melas*, a kind of dolphin.

It was called *deductor* in the nineteenth century because of its habit of following a leader. Shetlanders liked to eat them. 'They are harmless, inoffensive creatures,' thought Frank, 'and I should imagine excessively stupid, as they are continually running ashore, allowing themselves to be murdered wholesale.' Identifying it was one thing, but making a cast was quite another. Frank set aside Saturday afternoon and ordered 'a vanload of plaster'. The size of the creature meant the cast had to be made in six pieces. Frank's crew of volunteers cannot have known what they were in for, but neither could Frank. Until he tried to make a mould from it he could have had no idea that the rascally supplier had sent him plaster of such poor quality that it would crumble like icing when it dried. This meant that a second batch had to be mixed and slapped on over the first. The problem now was weight. The reinforced mould had 'the thickness of armour-plating on an ironclad' and was massively heavy. The head and sides came off cleanly enough, but then . ... 'when attempting to turn over the largest portion of the mould, about 9 feet long, one side, without the least warning, gave way in the most aggravating manner, and crumbled like a piece of pie-crust'. Anyone who thought Frank would now admit defeat and send his helpers home was swiftly disabused. He wasn't going to be beaten by a mess of faulty plaster, any more than he would be beaten by a faulty salmon ladder. With 'difficulty and much contriving', they managed somehow to stick the broken pieces back together. Frank allowed his flagging crew a short rest before driving them on to make a cast from the mould. He sat down to dinner at five minutes to midnight, and the cast duly made its way to South Kensington. The point of this story is not that it was in any way exceptional. In the organised chaos of Frank Buckland's life, it was just another day: one more problem solved, one step closer to exhaustion.

It is a moot point whether the most challenging of his labours was inspecting rivers or collecting ova, which he continued to supply on behalf of the Acclimatisation Society to rivers across England. This account of his 'spawning kit' shows the discomforts he faced, and the load he had to carry:

First, the waterproof dress; this very useful garment is in fact a diver's dress, and when properly put on, admits not a drop of water. It has however one fault; it is apt to freeze when I am out of the water, and then one feels encased as it were in a suit of inflexible tin armour. Second, the spawning tins: long experience has shown me that the best form of vessel in which to spawn fish is no paltry little tin, but a bath large enough to bathe a good-sized baby in. Third, a long shallow basket, to hold the salmon immediately they are caught, such as ladies use to carry their clothes when travelling; such a basket makes a capital salmon cage. It is easily sunk to the proper depth with stones, and the salmon live well and do not fret in it. The fish can remain alive in the water in this till they are wanted. I find a big 'crinoline net' does not answer, for the fish are apt to turn on their sides, and get faint; besides, there is a difficulty in getting them in and out. Fourth, house flannel cut into lengths of one yard; this is absolutely necessary to hold the struggling salmon. Those who are unaccustomed to spawn salmon, have an awkward habit of putting their fingers into the gills of the fish, and if the fish's gills are injured and bleed he suffers much from it. I never to my knowledge killed a fish in my life while spawning it. Fifth, dry towels; these are most necessary, as the slime from the salmon makes one's hands very slippery, and it is a very bad thing to have slippery hands, for you may lose a valuable fish in a moment; besides which, wiping the hands warms them, and when working in the water at this time of the year the cold to the hands and arms is fearful, or, to use the expression of the cabmen, 'all my fingers is thumbs'. Sixth, bottles for experiments with milt and ova. Seventh, 'Sphagnum' moss to pack the eggs in the tins and cans. Eighth, wooden boxes tied up in threes and fours to pack the eggs at the river side. Ninth, a huge landing-net. This is very useful for catching fugitive fish, which may happen to escape round the side of the net. Tenth, nets; one long heavy trammel and two small trammels are best. Eleventh, ordinary luggage, and especially a bottle of scented hair oil with which to well anoint the chest and arms and tips of the ears when working in the water, a most excellent and serviceable plan. I took this hint from the Esquimeaux.

Oddly, he seems to have had a strange aversion to boots, which he would kick off whenever he wasn't actually walking in them. This could cause embarrassment. Spencer Walpole remembered an occasion when Frank fell asleep with his feet on the windowsill of a railway carriage. The inevitable happened, the boots disappeared onto the track and Frank arrived at his hotel in his socks. Such was his fame, however, that his eccentricities were known to all. The platelayer who picked up the boots passed them to a traffic manager, who learned that Frank Buckland had passed through. No more questions asked. The boots were dispatched to the Home Office, where they soon resumed their on–off relationship with their owner.

The feet were seldom still. Two weeks' entries from Frank's diary of August 1867 show the intensity of his schedule. On the 19th he set off to visit the Duke of Northumberland at Alnwick, taking the ten o'clock train from London and writing all the way to Newcastle. Next morning he was up early to inspect a weir, then met local conservators; 'after that dragged the river, and then on to Alnmouth'. The 21st was even busier. 'To see the artificial hatching place; with post horses in Duke's carriage, to Wentsworth Mill, and then back to Alnwick. By train that night to Newcastle.' A bit of spare time next morning found him walking around Newcastle, visiting the fish market and the museum. Then it was on to meet conservators at Hexham, and on again to Carlisle. On the 23rd: 'Drove all down the Aln. Meeting at Maryport.' The appointment was with mill owners, whom he disarmed in his usual fashion. On the following day he returned to London. There was no entry for the 25th, which was Sunday and would have seen him at his regular place of worship, St Mary Magdalene in Munster Square, where he and Hannah had been married. Next day he was on the train to Chepstow, where he inspected the River Wye. On the 27th it was the River Towey at Carmarthen, 'then on to Conyd'. On the 28th, he rose early to see the 'falls of Penarth'* before continuing onward to Newcastle Emlyn and Cardigan.

---

* It's not clear what he meant by this. Penarth lies on the north side of the Severn Estuary at the south end of Cardiff Bay. It is well known for cliff rockfalls, and it may have been these that Frank wanted to see.

On the same day: 'met Inspector of Police at Newcastle [that is Newcastle Emlyn, a town on the River Teifi straddling the border of Ceredigion and Carmarthenshire, not Newcastle-upon-Tyne], and ordered a new pass for the fish'. On the 29th he was back on the Teifi at Cardigan, whence he drove at night to Newport. The 30th began with a five o'clock bus via Fishguard to Haverfordwest. 'Took fly there; looked at two weirs on the Cleddy; and also paper-mills; then on to Langham, and thence to New Milford.' On Saturday the 31st he moved on to Pembroke and took a train from there to Chepstow, from where presumably he returned to London. His verdict: 'a very pleasant week'. He did the same in every season, in all weathers, putting his body – and Spencer Walpole's – under more stress (I would guess) than any other bureaucrat in the entire history of government.

All the time he was learning. Any trout or salmon that came his way would be stripped down to its component parts. He knew what they ate because he looked into their stomachs. I have learned from long and chastening experience to be wary of superlatives – *biggests*, *oldests*, *firsts*. Was Frank the first man to work out that salmon navigate by sense of smell? I don't know, but I've found no evidence to suggest otherwise. He was certainly far ahead of his time in proposing a theory that remained unproven until the 1950s. There is still some mystery about how salmon find their way from the deep ocean back to their birthplace. For the first stage of the journey, from ocean to estuary, scientists have suggested some kind of 'map and compass' system in which the fish make use of environmental cues – hours of daylight, position of the sun, the earth's magnetic field, changes in water temperature – to fix their direction of travel. So far so mysterious; but science now has a much clearer idea of how salmon manage the second stage of their journey, from estuary to the tributaries of their birth. It was shown by experiment in the 1950s that the young salmon, or 'smolts', become sensitised to the particular combinations of odour on their journeys downstream from the hatchery to the sea. These odours, never forgotten, form an olfactory map which guides them home again.

Frank also had some thoughts about raising fish in ponds. The better the fish were fed, he counselled, the faster they would grow, and he knew exactly how to do it: 'I strongly recommend that a dead cat or rabbit, unskinned, should be hung up in a tree over the pond. The gentles [maggots] resulting from the blow-flies will fall into the pond and afford excellent food for the fish.' It probably helped if the pond's owner lived

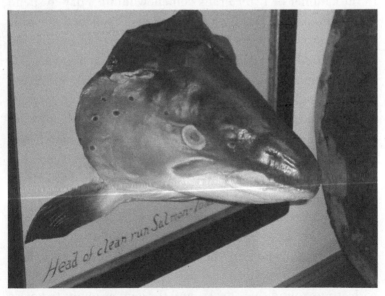

Frank's cast of a salmon head, in the collection of the Scottish Fisheries Museum

upwind. With the same breezy insensitivity, Frank warned against overstocking.

> One thing is quite certain, that the more numerous the fish are the smaller they get. Nobody ever yet saw a fat charity boy. Charity boys and girls are generally about the same size, and run thin; the reason is, that they have just enough food to keep them in health, but not enough to make them fat – the guardians take good care of that. The same rule that applies to charity children applies to fish.

Frank always enjoyed the company of fellow enthusiasts. One of these was Captain Francis Henry Salvin, a contributor to *The Field* whose passion was fishing with cormorants. In the summer of 1867 Frank somehow found a day to spend with him on the River Test, where Salvin launched a couple of birds into a mill pool. These quickly latched onto a fish and, as Frank put it, joined with it in a 'regular set-to, like a steam-engine had gone mad at the bottom of the river'. Standing and watching, however, was not Frank's forte. 'I could stand it no longer, so I jumped in the water, and went to the assistance of the birds.' The fish had squirmed under some boards, where it was out of the cormorants' reach but not out of the stout gentleman's. After a great deal of splashing and hallooing, Frank flipped a 3lb grayling onto the bank, where 'our friend Salvin greeted us with loud shouts, and vehement cracks of his cormorant whip'. Slightly less pleased was the miller, 'who said "he never see'd three such poachers in his life in the water together, as them long-necked birds with straps round their necks, and the gent as ought to have a strap round his'n" '.

In *Land and Water* Frank amused himself by gently putting right his readers' misconceptions. One of his most persistent correspondents was a Mr Gilbanks, who somehow got it into his head that the best way to control beetles in a house was by letting loose a hedgehog. Frank's humour was one of the principal reasons why the reading public loved him. He was clearly fond of Mr Gilbanks.

> [He] said at a public dinner the other day at Maryport that 'he and I had had many a passage at arms, but that we were none the worse friends for that'. I now, therefore, again call on the squire to call his sword. Protocols have failed, and it is a *casus belli in re hedgehogs*. I have tried hedgehogs to kill beetles; they don't act. A hedgehog cannot possibly hold above a pint of beetles at a time, and in my kitchen there are gallons of them.

At the Westminster Deanery, an imported hedgehog had not only failed to deal with the beetles but also found various other ways to make itself a nuisance, stealing soup and rattling the silver at night in a perfect impression of a burglar. The animal soon afterwards found its way into a flue, where it compounded its felonies by smoking the house out. Mr Gilbanks meanwhile, to Frank's great delight, was compounding his errors. Frank wrote chidingly:

> Again, good squire, who told you that tortoises ate beetles? Surely no-one but the London costermonger who wished to sell you one. We might as well shut up the squire himself in his own kitchen to eat blackbeetles, as shut up a tortoise to perform that task. If my statement is doubted, shut up a tortoise with a dozen black beetles: I warrant the dozen

beetles will remain there till the tortoise dies of starvation, and it is then exceedingly likely that the beetles will eat up the tortoise.

Having already shown that a hedgehog would eat a grass snake, Frank now wanted to know how it would deal with a viper. His observations were as brisk as the encounter he described: 'The viper struck the hedgehog two or three times in the face, where there were no bristles; the blows were well aimed, and meant to do business, as at the moment the hedgehog was munching up the viper's tail. The hedgehog did not suffer in the least; on the contrary, he ate up the viper in a few minutes, leaving not a trace behind.'

Unknowingly, he was an early proponent of the tabloid newspaper maxim: *listen to the nutters*. When someone told him, for example, that salmon in the Seame leaped in response to church bells he did not simply dismiss it. He thought about it. And in a way it was true. Salmon did indeed become active when the church bells rang at eleven o'clock on Sunday mornings. Why was this? Could bells really make fish jump? *No*: the actual stimulus was the extra volume of water coming down over the weirs when the mills were shut on Sundays. 'The salmon find this out, and, like wise fish, make the best of their time in endeavouring to get over the weir.' But fish at weirs were not always a happy sight. Frank described what he had seen at 'a terrible obstruction called the Acklington Dam' on the River Coquet, a scene that 'really made my heart ache'. The dam was a curved wall, eleven feet high, impossible for the fish to leap over, though they tried. How they tried! In the space of a minute, Frank counted twenty-seven salmon hurling themselves at the barrier and tumbling back again. They were magnificent in their futility.

The fish came out of [the deep pool below the dam] into the air with the velocity of an arrow; they gave no warning or notice of their intentions, but up they came, and darted out of the surface of the water with a sudden rush, like rockets let loose from the darkness of the night into the space above. When they first appeared in the air, their tails were going with the velocity of a watch-spring just broken, and the whole body, sparkling as though enamelled, was quivering with the exertion ... As they ascended, their tails left off quivering, for these tails were machines made to act on water, and not wings to act on air ... Not one fish, alas! did I see get over; some of them jumped into the body of the waterfall, and were hurled violently back, down into the pool, like the pictures we see of soldiers of old being thrown down headlong from the ramparts of a besieged city.

For some, the margin of failure was agonisingly small. Frank as usual wanted accurate data, and would risk life and limb to get it. Up he went onto the slippery wall of the dam, armed with a measuring tape. The bigger, more powerful salmon, he found, were jumping nine feet vertically into the air, a superlative feat that deserved better than to be hurled back, bruised and half stunned, into the pool. Frank pinned up a notice, ostensibly for the attention of the fish themselves but more likely designed to prick the conscience of the Duke of Northumberland, who had yet to honour his promise of a fish ladder.

NOTICE TO SALMON AND BULL-TROUT
No road at present over this weir. Go down stream, take the first turn to the right, and you will find good travelling water upstream, and no jumping required. –
F.T.B.

It was a reminder also to himself, of the scale of the task that lay ahead, one dam after another, river after river. The fourteen-hour days stretching ahead of him were a sentence for life. For him there would be no reprieve.

CHAPTER THIRTEEN

A Pound of Flesh

It was not just at Epsom that Frank took an interest in horse-flesh. Inspired perhaps by the French, he saw horse as the obvious way to feed the hungry masses. Britain swarmed with horses of every kind from miniature ponies to leviathans of the plough. They all had to die (many of them very young from over-work), so what made better sense than to recycle them as steaks and stews? It was with great optimism, therefore, that London's intellectual elite gathered at the Langham Hotel on 6 February 1868, to set a public example with a grand horseflesh dinner. Frank only wished he had the skill of a Hogarth to record the diners' faces.

Many . . . reminded me of the attitude of a person about to take a pill and draught; not a rush at the food, but a 'one, two, three!' expression about them, coupled not infrequently by calling in the aid of olfactory powers, reminding one of the short and doubtful sniff, that a domestic puss not over-hungry takes of a bit of bread and butter. The bolder experimenters gulped down the meat, and instantly followed it with a draught of champagne, then came another mouthful and then, as we say, *fiat haustus ut antea*.

We can't now know whether the meat was moorland pony, tender young foal or worn-out carriage horse. All we do know is that it was served in a variety of different ways and that, in Frank's opinion, it tasted like 'the aroma of a horse in sweat'. This might have surprised as well as disappointed him, for he was no stranger to what he called *hippophagotomy*. Indeed, his enthusiasm had been piqued by a lunch with Abraham Dee Bartlett at which they blind-tasted 'two exceedingly fine hot steaks'. One of these was 'rump-steak proper', and the other a slab of horse. Both of them had preferred the latter. 'Uncommon good' was Frank's verdict.

He concluded not unreasonably that the quality of the meat was determined by the quality of the animal. With species bred for slaughter – sheep, cattle, pigs – this was easy enough to manage; but the horsemeat trade, then as now, was the province of rogues. Today in Britain horse appears only when illegally masquerading as beef in 'economy' burgers and pies. An epidemic of adulterated meat products in 2013 triggered the most powerful surge of public emotion in England since the death of Diana, Princess of Wales. And so it was in the nineteenth century. Not even the dire shortage of affordable meat could transcend the public's revulsion at putting horse in their mouths. It gave Frank a novel idea for deterring crime. All that was needed to keep men straight, he argued, was a diet of horsemeat in Her Majesty's prisons. 'Be assured these fellows who would garrote you, murder your wives and children, or commit the most fearful crimes, would shudder at the thought of dining upon horseflesh.'

The truth was that horse was knackers' meat, good only for the cat's meat men. Henry Mayhew took a close look at these lowly tradesmen in *London Labour and the London Poor*. There were hundreds of them in London, hawking the remains of the nine hundred or so horses that died and were boiled down in the city every week. The meat was sold at twopence-halfpenny a pound, or in small pieces on skewers for a farthing. Competition was fierce and unscrupulous. According to Mayhew, men would knock at houses regularly served by their rivals and offer cheaper deals, though it was a foolish customer who took the weight on trust. A cut-price 'pound' of flesh was often a sly twelve ounces, though when it came to sharp practice the tradesmen didn't have it all their own way: customers could be just as crafty. 'Old maids' in particular were notorious for unpaid debts. It was the cruel reality of this trade – heavy horses flogged so badly that after two years' work they were fit only for leather, glue and pet food – that inspired Anna Sewell to write *Black Beauty*. Frank himself was a regular customer, which gave him the opportunity to observe something very like the 'classical conditioning', or Pavlovian response, that would bring fame to the Russian physiologist Ivan Pavlov in the 1890s. Pavlov's dogs learned to associate food with the ringing of a bell; Frank's cats learned to run downstairs at half past one each day when they heard the cry of 'meat'. The response of Jemmy, Frank's pet meerkat, was to slink down behind them and wait its chance. Frank wrote a scratch-by-scratch account of the fight that usually followed.

The cat, annoyed by being disturbed at dinner, would leave off eating and strike sharply at Jemmy with her paw; that was his opportunity. In a moment he would seize the cat's meat and bolt with it, but by a most ingenious method, for when within striking distance of the cat's paw he would turn round and back up to the cat's face, and directly she struck at him, he caught the blow on his back; then he would put his nose down through his fore-legs, and through the

hinder ones, and have the meat in a moment, leaving the
cat wondering where it was gone.

The story ended with a moral caution. Jemmy's penalty was to
grow 'as fat as a little bacon pig' and die of a fit caused, so his
grieving master believed, by overeating.

Near Herne Bay in 1866 Frank had encountered the misleadingly
named Judy Downes, an elderly fisherman (real name probably
Judah) who earned his living from fish stranded by the tide in a
weir on the shore at Hampton. His haul one morning included
a stingray, which Frank bought from him before he could chop
off the murderous tail. Frank sensibly treated his trophy with
caution, leaving it to thrash away in the mud until it died. Even
then he did not much like the look of it. '[Its face] is like that of
a very hideous baby in the act of perpetual crying. His teeth are
very small, as though nature had taken out the two largest teeth
from his mouth and stuck them into his tail. My friend has since
been immortalised in plaster-of-paris.' What grieved him about
Judy's enterprise was the waste. 'As the tide left the weir, I saw
a sight that made my piscatorial heart sad, viz., the destruction
of such quantities of young fry . . . The lower part of the weir
was quite silvered with – think of it, ye epicures! with whitebait, pretty silver and green, large-eyed whitebait, flapping their
innocent tails in the mud, gasping for water, breathing only the
fatal air.' He asked Judy what he did with them. Answer: *nothing*.
Aside from the odd bucketful used as bait for eels, he 'just let
'em lay'. Frank implored him to be silent. 'Say no more, Judy,
I beg, my feelings will not bear it.' It was not just the whitebait
that appalled him; it was the loss of the adult fish they might have
become. In a single bucketful he identified seven different species
including sole, flounder and plaice. Sadness having robbed him

At the peak of his powers: Frank photographed at Elgin during his inspection of Scottish salmon fisheries, June 1870

(only temporarily, let us be glad) of his taste for hyperbole, he could only state the obvious: 'The destruction weirs of this kind must do, if greatly multiplied along the mouths of rivers, must be very great.' It was a thought that would bear heavily on him for the rest of his life.

In 1869 Frank was one of two inspectors appointed to report on salmon fisheries in Scotland. The other was a Scottish official, Archibald Young, representing the Commissioners of Scotch Salmon Fisheries. Together in May, August and September 1870 they inspected more than sixty rivers. Frank then was in his physical prime, broad-shouldered and hefty. Given his penchant for immersion in cold water, this was just as well. In Bompas's account, Frank became 'almost amphibious; wading the pools below the weirs, feeling the force and direction of the current, and striving, so far as is possible to man, to enter into the feelings of a salmon'. Archibald Young's respect for his English colleague was warmly expressed in the obituary he would write years later for the *Scotsman*. Frank, he said, was a man of 'kind, genial and obliging disposition . . . He was about the most true and genuine man we ever met, without a particle of affectation . . . and it would have been difficult to say whether he was most at home in the polished society of a luxurious country house, or while engaged in demonstrating the anatomy of a salmon, a herring, or a lobster to a group of fishermen assembled round a fishing-boat on the beach.'

The affection was reciprocated. On the train back to London Frank wrote down his 'Impressions of Scotland', much as an anthropologist might record his impressions of a South Sea island. 'The people are kind-hearted, and most hospitable, and, above all, highly intelligent; everybody seems to be educated. In all the towns and villages I visited, regularly every morning the streets were crowded with children going to school at 9 o'clock, not creeping along according to the old adage . . . but stepping along with an active and light gait, as though learning was worth having.' Odd as it now seems, he applauded the fact that most of this 'stepping along' was done without shoes or stockings. 'An excellent custom; it saves expense, and makes them healthy. The more you walk upon shoe leather the thinner it gets, the more you walk upon human leather the thicker it gets.' He examined the feet of some fisher girls at Findhorn and found their soles 'as hard and as thick as the foot of an elephant'. He felt the usual Englishman's surprise at the rarity of kilts – in more than a month he saw only nine. 'At Nairn they told me, that, if you saw a man in a kilt in Scotland, you might be quite sure he was an Englishman.' Around Peterhead, he noted, 'the peasants' had peculiar wrinkles like crow's feet round their eyes, which they kept half shut, 'after the manner of Esquimaux'. A woman there told him the wrinkles resulted from generation after generation of people puckering their eyes against the snow in winter and the dust in summer. It gave him an excuse for another swipe at the evolutionists. 'Here, then, is a crow's (foot) to pick for Mr. Darwin, a case of cause and effect.' Why he should have thought cause and effect inimical to Darwinian theory, rather than essential to it, is another of the mysteries he left us to ponder. Mysterious to Frank was the Scots' failure to appreciate the abundant local plovers' eggs (in London worth threepence each), and he did not enjoy the monotonous diet. '[It] was my lot to have haddocks in some shape or form for breakfast, and generally also for lunch and dinner every day.' He did, however, enjoy the earthy vigour of the women who gutted, cured and sold them.

I suddenly came upon twenty or thirty old fish-wives; they were sitting around a corner on the sand warming them-selves, and waiting for the boats . . . They were all knitting and all talking; my appearance made them suddenly silent; however, we soon made friends. I told them I wanted a Scotch fish-wife for my museum; 'Who will come?' said I. 'I will,' 'I will,' said all of them jumping up in a body. I was frightened at so many offers from so many fair ladies, and bolted like a shot. They all rushed after me shouting and holloing, and the oldest and ugliest very nearly caught me. But I was too quick, so she threw her fish creel after me, with a wild hooroo.

What upset him, as Archibald Young reflected, was the government's failure to accept the '23 carefully considered recommendations' in their report. Frank was forever dismayed by what he saw as British backwardness. England couldn't even bear comparison with the Chinese, who understood, for example, that there were better uses for sewage than polluting rivers. Treated as fertiliser, he said, England's sewage would grow wheat sufficient to make 405 million loaves of bread, 'enough to nourish five millions of people for a year'. What great nonsense it was to spend thousands of pounds on guano from the South Sea Islands while wasting what we could have for free. The answer was as plain as a call to nature: 'Take the pollutions from the river, where they are doing harm, and place them on the land, where they would be doing good.' Even in the twenty-first century this commonsensical approach still arouses suspicion in the public and caution among supermarkets, but every year more than 840,000 tonnes of treated sewage sludge or 'biosolids', odourless, dry and safe,* is spread on Britain's fields. The European Union and

---

* The 840,000 tonnes is dry weight, equivalent to 3.5 million tonnes of untreated sludge. According to the European Commission there have been no reported cases of 'human, animal or crop contamination' caused by spreading it on fields.

the UK government have recognised it as the Best Practicable Environmental Option, and the 'best possible alignment with the principles of the waste hierarchy'. The bureaucratic jargon would be as alien to Frank's ears as it is to normal English-speakers now, but the sentiment would chime like a bell.

On 19 April 1870, Mr Leckey, of the Falcon Hotel, Gravesend, telegraphed Frank with the exciting news that a salmon had been caught in the Thames. Frank hurried to buy it (two pounds six shillings was the agreed price), then brought it back to the smoking room in the Athenaeum. The fish was a big one – twenty-three pounds in weight, thirty inches long – and had been caught in Gravesend Reach. He wrote a letter about it to *The Times*, a ray of hope flaring in his mind.

The question arises as to whether this is a Thames bred fish. It must have been hatched somewhere; the nearest salmon rivers to the Thames on the South are the Itchen, Test and Avon, in Hants; and on the north the rivers falling into the Humber. I do not think this fish came from any of these rivers. I am afraid to venture a hope that it is one of the salmon which Mr. Ponder and myself have been breeding and turning out for the Thames Angling Preservation Society during the last seven years. This may possibly be the case, except that its history in the grilse state is unknown. Finally, it may have come across the Channel from the Rhine, for it has the general appearance of a Dutch fish, except about the head. Anyhow, I am glad to be able to report its arrival in the Thames, as it is a good omen for the future salmonization of this noble river, though, doubtless, it will be called by some (as its predecessor three years ago was profanely called) 'the annual nuisance'. The fish will be on view

at my museum of economic fish culture, Horticultural
Gardens, Kensington, during Thursday next.

He signed himself 'Frank Buckland, Inspector of Salmon
Fisheries, 4, Old Palace-yard, Westminster'. The 'salmonization
of this noble river' must rank as one of his most heroic failures.
It was always a tall order. Commercial shipping makes an
uneasy neighbour for clean-living fish, and no amount of
fry will make any difference. What is poisonous to the few
is poisonous to the many. As Einstein may or may not have
said, the definition of insanity is to do the same thing over
and over again and expect a different outcome. It has taken an
awfully long time for the penny to drop. Between the late 1970s
and early 2000s, upwards of £3 million was spent on attempts
to restock the Thames. It would have made just as much sense
to pour banknotes into Teddington Lock. In 2011, researchers
from Exeter University, funded by the European Union and the
UK Environment Agency, concluded that 'habitat reconstruction' – i.e. creating an environment which would attract fish
naturally – would be far more effective than restocking from
hatcheries. As it was, any salmon seen in the Thames were more
likely to be strays from other rivers than the 'Thames bred fish'
of Frank's wishful thinking. Individual salmon are occasionally spotted in the Thames, when they are invariably hailed as
piscatorial messiahs, but the Environment Agency's record of
rod catches on the river tells its own story. With one exception,
every year from 2003 to 2013 clocked a grand total of zero. The
exception was 2010, when there were two. I don't know what
Frank would have made of this, but I can't see him ignoring
clear evidence or allowing sense to be trumped by sentiment.
When funding is tight – i.e. *always* – conservation is the art of
the possible. Every pound, dollar or euro has to count.

Nevertheless, in 1870 Frank's salmon was a celebrity. These are his diary entries for four days in April:

> 20th – To Windsor with the salmon. Down the Thames in steamboat of Board of Conservators. Called on Prince Christian at Frogmore, and showed him the salmon.
> 21st – Sent Salmon off to the Museum. Neville tells me an enormous number of people came to see the salmon.
> 23rd – Cast the Thames salmon five times. Took one of the casts down to the Royal Society.
> 29th – Sir R. Murchison's *soiree*. Took down casts of group of lizards and puff adder, and the Thames salmon; all much admired.

*

Albany Street continued to be the destination of choice for anything deemed by its sender to be bizarre, dangerous or edible. In April 1868 Frank took delivery of a steak from a bull bison which Lord Wharncliffe, of Wortley Hall, Sheffield, had sentenced to death for crimes of violence. 'Grand eating,' said Frank, 'many degrees better than horse-flesh.' Other parcels could be trickier. Late one night Frank came home to find a jeweller's box lying open on the table. Inside it was a scorpion with its sting cocked – enough to startle even him, though not as much as it had already startled the household. The joke was on Hannah – the 'custom-house officer', as Frank called her – whose curiosity about the seductive-looking, gold-lettered box had overcome her discretion. When she opened it, 'there was a scene rivalling that when Blue Beard's wife opened the mystic cupboard'.[*] Box and

---

[*] In the classic fairy tale, she found the strangled bodies of Bluebeard's former wives.

scorpion were hurled into the grate (happily not lit), whence they were recovered by the butler, John, a man trained to expect the unexpected, who was alerted by Hannah's shrieks but relieved to find the scorpion was dead.

A home like no other: 37 Albany Street

In February 1869 the postman delivered two more scorpions, live ones this time, sent by the entomologist J. K. Lord, who had caught them under a stone in Egypt. Frank's fascination with all things venomous meant they had to be tested. First he matched them against each other. The arena was a glass fish bowl in which they advanced head-on and locked claws 'like two bulldogs'. Result: stalemate. Frank prised them apart with forceps and replaced one of them with a live mouse. This was much more interesting. The mouse struck first, biting the scorpion on the back before receiving in return a sting between the ears. The combatants then locked together and rolled over, 'like two cats fighting', while the scorpion went on stabbing 'with the velocity . . . of a sewing machine'. The mouse countered by pinning the scorpion's tail with its paw, nipping the sting with its teeth and biting off two of the scorpion's legs before retiring to the corner 'to wash his face and comb his fur'. Frank watched to see what effect the poison would have, but the mouse seemed nothing worse than a trifle weary. Frank nudged the scorpion back into the fight, but it was too far gone. It threw in the towel, and over the next few hours was gradually swallowed by the mouse, helped down by a morsel of bread. The winner duly received his accolade. 'As a reward for his courage, the mouse, after a parting supper of toasted cheese and milk, was let free in a place where the cat was not likely to find him.' Yet Frank conceded that 'the fight was hardly fair, as the scorpion had just come off a long cold journey, and had not eaten anything. The mouse, on the contrary, was fresh caught and in good fighting condition.' Verdict: technical knockout only. Judgement reserved.

Another unsolicited gift was an eight-foot tuna. This had died in a pilchard net at Dawlish in south Devon, but none too recently. Its flesh was 'red as beef', bloody and tough-looking, but it seems to have been too high even for Frank's ironclad constitution. What an irony, after all his experiments with elephant, giraffe, panther and wombat, and all his hopes for bison and eland, if he never sampled what would become one of the next century's most ubiquitous foods.

His liking for big fish meant that letters and telegrams rained down from all over Britain and beyond: a thresher shark at Brighton, another shark (species not specified) at Dover, a swordfish in Cornwall, an enormous fish, twenty-seven feet long and probably a basking shark, at Achill Island, off County Mayo. And all the time he was spreading the word: a paper on oysters, mussels and deep-sea fishing for a Society of Arts inquiry into 'the supply of food for the people'; a lecture on salmon and oyster culture for the British Association at Norwich; lectures on salmon at Gloucester, Birmingham, Falmouth, Leamington Spa, Herne Bay and Holloway. None of this got in the way of his inspections of rivers and weirs, which proceeded unabated, while he continued his profuse literary output and went on casting specimens for the museum.

One of the strangest calls on his time came in December 1868 when he was summoned to give evidence in what *The Times* described as a 'very curious' case, *De Garris* v. *The Mercantile Marine Insurance Company*, heard before the Lord Chief Justice in the Court of Pleas. The case hinged on an incident at sea on the afternoon of 13 March 1864, when a sailor on the barque *Dreadnought*, en route from Colombo to England, hooked a swordfish so powerful that it broke his tackle and escaped. Soon afterwards the ship sprang a leak. What the court had to decide was whether or not the hole in the planking had been made by the swordfish. If it had, then the plaintiff would have a claim against Mercantile Marine, with whom the cargo was insured against damage caused by 'contact with some substance other than water' (e.g. an angry swordfish). The defence not surprisingly challenged this. Counsel ridiculed the idea that 'a fish with a shark hook in its mouth, with 25 yards of line attached to it, had such affection for the vessel as to return to it, having once escaped'.

The court wanted expert advice on the likelihood of a swordfish being able, first, to pierce the planking, and then to withdraw its sword and swim away. Frank and Richard Owen were called as witnesses. On the first point they were agreed. Swordfish had been known to pierce ships to a depth of eighteen inches or more. In every recorded instance, however, the sword had broken off in the hole. Crucially in the *Dreadnought* case no sword had been found, and here Owen prevaricated. As only three inches of wood had been pierced, 'he was not prepared to say that the fish could not have got out'. Frank was bolder. *The Times* summarised his evidence thus:

> Mr. Buckland said a swordfish could go forwards, but not with any force backwards; he could not reverse his action. He thought it was very unlikely, if a swordfish had pierced the Dreadnought, that he would have been able to get free again. He had read every known book on swordfishes. Plenty got in, but none got out . . . [The fish] might make a great fuss laterally, and might so wriggle out, but he was more likely to break his beak off and kill himself.

Frank added that he had noticed when handling live salmon that they always fought to swim forwards, never back. 'But', objected counsel for the plaintiff, 'you never had hold of a live swordfish by the beak and tried whether you could prevent him from going backwards?' 'No,' said Frank, 'but I should like to.' The jury found for the plaintiff.

That was not the end of Frank's interest. With his usual reluctance to let a subject drop, he wanted to discover all he could about the fish and their swords ('I find they make capital paper-knives for newspapers'). He found that the market price for swordfish at Constantinople was the equivalent in piastres of one shilling, ninepence-halfpenny – 'rather dear for this kind of fish, I should imagine' – but that they were good to eat. He believed the sword to be hollow, and proved it by puffing cigar smoke through one. This led him to propose what might seem an unlikely, not to say preposterous, comparison with the bill of a duck. The sword, he suggested, contained 'nerves of sensation' which guided it when 'rout[ing] up the mud and sand for his food'. The duck's bill was very similar.

> This bird searches for minute creatures in the mud, and if the reader will take the trouble to dissect off the skin from the upper beak . . . he will find a large white nerve on each side coming out of the skull, just in front of the eye, and ramifying itself on the tip of the bill, so that the bird has, as it were, the tip of a finger on the end of its bill, and by the power of touch can select proper, and reject improper food. In my humble opinion, nature always economises material to its utmost extent, and every organ in the structure of every animal has its use, *if we could but known what that use is*. This is the doctrine of Teleology: i.e. the doctrine that every organ is adapted to a special use.

His use of *adapted* here is interesting. For Darwinians, adaptation is evidence of evolution. To Frank, it proved the perfection of God's Creation. In the case of the swordfish, however, he seems to have misread the Creator's mind. Goodness knows what made him think swordfish were sand-stirring bottom feeders. In fact they feed close to the surface, where they swallow small fish whole and slash (but not stab) bigger ones with the sword, which is a bony extension of the upper jaw. In a scientific age we might now deplore Frank's tendency to rush to judgement. But enquiring minds in the nineteenth century had little to rely on but instinct, imagination and, in Frank's case, faith. There were no well-funded research teams tracking animal movements and behaviour; no camera traps; no opportunity for underwater observation. Much of nature remained as mysterious, and as invisible, as God Himself. The importance of men like Frank was not so much that they etched their truths in stone, it was rather that, as far as their faith allowed, they were champions of rational enquiry, fitting their conclusions within a framework of fact and changing their minds when more was known. Though Frank was unshakeable in his faith, he never asserted any opinion for which he could see no justification in fact. When he was wrong, it was somehow in the right way. More often than not his instincts were correct. Further evidence of that came in the same month as *De Garris* v. *The Mercantile Marine Company* when the Australian agriculturalist James Arndell Youl published a letter in *The Times* describing the exportation of salmon ova to New Zealand and extolling Frank's part in the enterprise – not least 'a present of 5,000 beautiful sea-trout ova, which he himself had manipulated' and which he had stowed in the ice house aboard the ship which carried them to Otago.

The balance of Frank's mind, suspended in the most delicate of equilibriums between fact and faith, was spelled out in his journal. His mind, he considered while reflecting on his forty-second birthday, was 'hardening into an adult mind' capable of deep study across the wide range of issues that his work as fishery inspector involved him in – business, 'conduct of affairs' and legislation as well as natural history. Like most men of learning, he floated between arrogance and humility. He didn't doubt the scale of the task that lay ahead, but neither did he doubt his ability to tackle it.

> I must teach myself, as I have had to teach myself nearly all I know: but, thank God, the Dean gave me a good soil at Oxford, which will grow almost any seed placed in it, and I must now plant a new seed. Those of observation have grown into big trees long ago; I must now plant the tree of mental reasoning upon things observed. I am truly thankful for all mercies, because all is the gift of God.

In 1869 Frank's exhibits at an Exposition of Fisheries at Le Havre earned him a Diploma of Honour, a distinction he shared with the chief engineer at the French Etablissement de Pisciculture and the King of Siam. He also received a medal from the Acclimatisation Society of Victoria and a silver claret jug from the provincial government of Otago, both in gratitude for his gifts of salmonids to the Antipodes. He had his eye on other frontiers too. In November 1869 he sent models of salmon ladders, a photograph of his museum and five boxes of oysters to the American Fishery Commissioners in New York.

In September that year he enjoyed a starring role in an uproariously popular spectacle. The occasion was the draining of the Serpentine in Hyde Park. Work began on 24 September, though it was not until 11 October that the water had gone down far enough for the fish to be netted, and an elderly fisherman from Hammersmith was hired to trawl from a rowing boat. A huge crowd gathered to watch the fun. 'It is very extraordinary', Frank complained, 'how a London mob finds out when anything is going on . . . by the time we had got the net on shore so many people had assembled . . . that it was almost impossible to do any work at all.' To the dismay of the many small boys ('the gents is killing all the tittlers!'), the first sweep yielded only a few small roach and a shoal of sticklebacks. Such was the pressure of the crowd that the centre of operations had to shift to a railed embankment at the west end of the lake where it was cordoned off by police. The next trawl almost ended in tragedy. Hauling at his net, the fisherman stepped onto what looked like solid ground but which was in fact a six-foot depth of near-liquid mud. A human chain, Frank at the forefront, linked hands to drag him out. When the old man finally popped out like a cork from a bottle, the crowd cheered and laughed as if it were at a music hall. Next time the net was hauled in, Frank reached into the seething mud and held up a large bream like an executioner brandishing the head of a traitor. 'The crowd . . . gave him a great shout of welcome which would have done honour to a whale.'

The next challenge was to transfer the fish by watering cart to the Round Pond in Kensington Gardens. For a while it all went well. After the first cartload – 'about 500 fish, little and big' – the nets were joined together for one last haul. Expectations were high, and the fisherman was not disappointed. There were enough fish to fill three water carts, which rumbled along with an escort of 'hundreds of dirty little vagabonds . . . wild with excitement'. Frank threw them a fish or two and watched as they fought over them 'like hounds breaking up a fox'. Then, disaster! A cart slipped on the bank, tipped end over end and catapulted a man called Lee* into the water. Frank seems to have relished the drama:

---

* Almost certainly the writer and naturalist Henry Lee, who would become naturalist manager and then director of the Brighton Aquarium.

he certainly enjoyed describing it: 'The great shafts came rushing down from above like two scaffold poles, and the body of the cart fell right over into the water, wheels uppermost . . . My first thought was for my friend Lee . . . something came down on him with a tremendous crash, and he disappeared clean under water. I rushed immediately into the water, nearly to a swimming point, and got hold of his collar: he came up in a second, blowing like a grampus.' While apparently philosophical about the accident – these things happen – Frank was outraged by the behaviour of the onlookers.

> The crowd . . . stood like a lot of marble statues; nobody offered to move, or say, or do, or suggest anything. Upon my word, I think an English crowd is either very selfish or exceedingly stupid. I think the people of the present day are a great deal too highly educated, as it is called. They are too clever to think or act quickly enough in a case of emergency. Competitive examinations will never teach presence of mind, or suggest a remedy for a sudden accident.

The final tally was eleven cartfuls, about 5,500 fish. Most were roach, though there were some bream, a few tiny Prussian and crucian carp, some tench, a deformed perch and an eel. There were no pike. The eel was a big one – six pounds in weight – which Frank cast for the museum, but it was literally ill luck for the taxidermist who stuffed the skin. He and his family ate the flesh and suffered hours of sickness as a result. Frank blamed the scarcity of eels, absence of perch and pike, and the overall smallness of the catch on the lime which had been put into the Serpentine some years earlier to purify the water. In man's edgy relationship with nature, the law of unintended consequences has a long and unhappy history.

It was not just fish that were netted. The mud gave up a drowned armada of model yachts and enough household relics to stock a museum. Up with the trawl came: 'Pickle bottles, wine bottles, soda-water bottles, blacking bottles, ink bottles, physic bottles, ginger-beer bottles, beer cans, sardine cases, coffee-pots, tea-cups, egg-cups, shoes, boots, pipes without end, dogs' bones, cat's heads [*sic*], skate straps, gallipots, a top of a lamp, india-rubber balls, cocoanut, curtain rings, a loaded shot cartridge, an iron weight . . .' Prize finds were a human skull and a sword made from a scythe blade, 'of the kind favoured by rioters'. Disappointingly, there were very few coins, a fact which Frank thought said something about the national character.

> If this had been the remains of a Roman camp, we should
> have been sure to find plenty of coins, but English people
> don't seem to throw their money away in the curious manner
> that appears to have been fashionable among the Romans.
> If we ploughed up an English camp, such as Aldershot, we
> should find that the English soldiers had dropped very few
> shillings or coppers; but the Roman soldiers . . . seem to
> have spread their money broadcast. I suppose they had no
> pockets in their armour.

The poor state of the Serpentine was nothing to set against Frank's despair at the rivers and coasts. He deplored the obduracy of those who either failed or refused to grasp the truth. If the nation wanted fish to eat, then it would have to make a better job of managing its stocks. An editorial in *Land and Water* in 1866 was one of the first shots fired in the ongoing battle between scientists and fishermen. To no country in the world, Frank believed, were the fruits of the sea more important than they were to Britain. Words such as *conservation* and *sustainability* had yet to enter the language, but they would have conveyed Frank's meaning precisely. From a state of plenty, he warned, we were set on a course for dearth. 'Cod, mackerel, herring, pilchard, salmon, oysters – the richest productions of sea and river, under favourable circumstances, are found in the greatest abundance. Yet we have permitted our shores to be denuded of breeding fish and fry by trawlers working too near inshore at certain seasons . . . The most erroneous ideas frequently prevail among our fishermen, displaying an amount of ignorance absolutely startling.'

He declared war on apathy. Not for the first time or the last, he contrasted English inertia with the forward-thinking attitudes of America and France, where international exhibitions were held to educate the public. England, he declared, urgently needed such an exhibition of its own. But it was another whistle in the dark: it would not happen in his lifetime. He tried another tack, first suggested in 1867 by a correspondent in *Land and Water* who wrote under the pseudonym Ptarmigan. Frank agreed with Ptarmigan that what the country needed was some aquaria big enough to hold large fish, where people could get a better idea of what went on under the sea. He also argued that the government should sponsor research at sea, 'in order that they may make a thoroughly practical examination of the dark and mysterious habits of food fishes'. The vessels used by revenue officers against smugglers, he suggested, would make perfect platforms from which to work. Again he had no luck. It would be many more years before government-funded research would become a reality, but he fared better with aquaria. It so happened that Eugenius Birch, the engineer famous for seaside piers (among them Margate Pier, Blackpool North Pier, Brighton West Pier, Aberystwyth Royal Pier and the piers at Deal, Lytham, Eastbourne, Scarborough and Hastings), had seen a small aquarium at an Exposition Internationale de Peche at Boulogne in 1866. This had given him the idea of building a much bigger one at the English seaside, and his eye had fallen on Brighton. Building began in 1869, though the aquarium did not open until 12 August 1872, a delay which allowed it to be overtaken by the Crystal Palace, which opened in January 1872 with speeches by Frank and Richard Owen. But it was at Brighton that Frank would be more closely involved. The aquarium's first general manager, John Keast Lord, did not long survive his appointment (he died in December of the same year), and Frank, along with Henry Lee and Abraham Dee Bartlett, was drafted in to help run it. Ironically, one of their co-directors was Frank's old enemy Francis Francis. From his unassailable position of eminence, Frank felt able to bury the hatchet. As he told readers of *Land and Water*: 'Though this gentleman and I have had many tiffs together on purely professional matters, still, in all fish matters which indicate progress and developed knowledge, I trust I may be excused for writing we are friends. Time is too short, and there is so much to be done that it's no use driving single harness any more.' I have found no record of what Francis Francis thought about this. One might wonder how Frank could possibly have found time to take on any more work, but he felt compelled to seize the opportunity and press his strongly held view that aquaria should be places for study and not just for amusement. As so often he was ahead of his time. Two years earlier, an editorial in *Land and Water* had put the case for a marine equivalent of his own role as Inspector of Salmon Fisheries.

What we want, and sooner or later must have, is an inspector of sea fisheries appointed by the Government, whose duty it should be to traverse the coast, sail in trawlers, and inspect the various systems of net and line fishing carried on at the different fishing stations, to take note of the arrival and departure of the migratory species; and to find out, if practicable, by well-conducted experiments, when and where the different kinds of sea fish we consume deposit their spawn. Concerning the sea we are groping our way in the dark, and every step we take is one of hesitancy, and thus we shall keep stumbling and blundering along until there are no fish left to catch, unless we at once grasp the lamp of science and guided by its light, boldly strive to find out for ourselves, what actually is going on amongst the fish down in daddy Neptune's diggings.

In his biography Burgess suggests that Frank must have been the inspiration for this, but he is anxious to acquit him of responsibility for the coy excrescence *daddy Neptune's diggings*. In this he saw the whimsical hand of Henry Lee, though Frank had his whimsical side too. Brighton, he declared, was 'a Solomon's temple among aquaria', the biggest and best that had ever been built.* He loved coming face-to-face with living fish. 'I fancied sometimes I saw a smile steal across the face of the largest cod. How do we know these fish have not a language? They don't speak, certainly, but they may converse by the eye.' Herring, too, fired his enthusiasm. 'I doubt whether any Brighton lassie could find a prettier object to put in her hat than a live herring.'

---

* Sealife Brighton, as it is now called, has outlasted all its contemporaries and is the oldest operational aquarium in the world.

Aquatic mammals found their way to the aquarium too, very often with Frank's help. In February 1873 a hamper left at Albany Street was found to contain two baby otters. They had been caught by Cornish fishermen who had killed the mother that had been trying to protect them – a seemingly weird but typically Victorian combination of brutality and sentiment. Their onward journey began with a cab ride to Victoria Station, where Frank kept the motherless babies warm in their basket on top of a gas fire, and where, after 'a bit of a shindy' with a porter who wanted to stow them in the luggage van, he embarked with them first class to Brighton. He did not record the reactions of fellow passengers to his constant ministrations with foot-warmers, a sealskin waistcoat and a drip-feed of sprats from his pocket. The otters rewarded him by escaping in the cab on the way to the aquarium and nipping his leg, resulting in a scuffle which Frank decisively concluded with a landing net. The triumph was short-lived. After settling the otters down with water and more sprats, Frank put them to keep warm by the furnace, 'where I left them as jolly as otters could be'. They died shortly afterwards. Nor were they the aquarium's only casualties. The porpoises died too, prompting Frank to offer some worldly advice to lovers. Forensic analysis showed the animals had been gassed. 'The last porpoise used to come to the surface and breathe, taking in a long inspiration [deep breath], like a spoony sweetheart when he leaves his lady love. The porpoise – spoony or not we cannot tell – sighed so deeply, that he blew out the gas light above his tank, and breathed a lot of gas into his lungs, and he never recovered it. Moral: don't sigh too deeply near a gaslight.'

In the spring of 1873 some live smolts were sent to the aquarium from the River Usk. The experiment began unpromisingly. All but one of them died. But then came the 'occasion of supreme felicity', as Frank put it. He doesn't say so, but we must conclude that the lucky survivor was kept in salt water. Hence Frank's jubilation at being able to announce that the smolt had become a grilse – i.e. a fish that had spent a winter at sea.

He is a wonderful and beautiful fish, the first smolt that ever turned himself into a grilse under the ken of us air-breathing mammalia. Naturalists can't live in the water, fish can't live in the air; so we make water cages for our fish, and we observe their wonderful transmutations from one stage of adolescence to another; changes quite as wonderful as the transformation of a dull-coloured, hairy vegetable-eating caterpillar, creeping along the ground, into a butterfly, which flies with ease in the air, on wings thinner than silver paper, and ornamented with colours, far more beautiful than anything that can be painted by artists.

We may note that Frank himself, for all his official grandeur and the accumulated wisdom of his forty-six years, had yet to complete his metamorphosis from boyhood to man. He never would. The boy was indestructible: the sprats in the pocket, the almost-naive capacity for wonder, the victory of excitable everyday language over the pedantic dirges of conventional learning. In a growing mountain of ironies, the habits that would so damage Frank's reputation were the very things that made him great. Who else would have taken time to wonder why messages in bottles were so rarely picked up? He favoured the theory of Admiral Hall, who blamed barnacles, which 'attach themselves to the bottles and breed so quickly that they sink the bottle altogether'. Another mystery solved.

The aquarium maximised its appeal with a sideline in popular entertainment. An article in the *Sussex County*

Magazine in 1952 recalled 'innumerable stage shows from serious drama to pantomime, and of such Victorian favourites as giants, midgets, Zulu chieftains, and Javanese temple dancers. In more serious vein it offered the greatest singers of the day, with Madame Patti* at their head, and an orchestra second to none.' No attraction, however, could beat Frank himself. In 1874 four hundred people paid 10s 6d each to travel by special train from London to hear him lecture and be led by him on a tour of the aquarium. They were also entertained to a lunch which might have been a little tougher on the jaws than some of them anticipated.

London Zoo's old rhinoceros had died. Frank had cut up the carcass and baked it in a pie.

* The operatic soprano Adelina Patti (1843–1919), described by Giuseppe Verdi as the finest singer who ever lived.

CHAPTER FOURTEEN

Ben, Glen and Mr Dog

After Brighton, aquaria became civic status symbols, ostentatious trophies for ambitious towns and resorts. In his capacity as national expert, Frank was heavily in demand. He spoke at the official opening of aquaria at Manchester in May 1874, and at Southport in September, each one eager to exploit his name. He was adamant that such places should earn their keep. Their proper function, he said, was as 'great educational schools, which will do much to teach kindness to animals, to humanise those but little brought into contact with the living works of the Creator'. Not only this: they offered vital opportunities for research. At his instigation they would study the life cycles of commercially important fish, and he continued to press the case for practicality. Students at Newcastle College of Physical Science, he said, should be looking for ways to improve their local fisheries, not wasting their time with books. Serious study meant getting wet.

Education was a drum he never stopped beating. He returned to it in the preface to *Log Book of a Fisherman and Zoologist*, published in 1875.

The so-called education of the present day is, in my opinion, too much confined to book-learning, and taking for granted the ideas and opinions of others. If I had my will

I would educate the eyes of all – adults even more than youths and girls – to observe and to photograph objects in their heads. I would also teach them to use their fingers to analyze and draw, and above all to dissect, Birds, Beasts and Fishes, so as to be able to understand their wonderful structure and mechanism . . .

I do not bow to many of the teachings of the modern school of science, which often, by hard words and unnecessary mystifications, frequently seems to puzzle rather than enlarge the mind. I wish, on the contrary, to throw the portals of Science, (*i.e.* knowledge) wide open, and let all enter who will; we want as many as possible in our ranks.

A hundred and forty years later the argument still rumbles on. Early in 2015 the Office of Qualifications and Examinations Regulation (Ofqual), which regulates the English examination system, stuck its fingers in its ears when both the government and the influential Wellcome Trust reacted furiously to its decision to drop lab work from GCSE and A-level examinations. The Department of Education expressed 'deep frustration'. *Plus ça change* . . . Frank provides no shortage of ironies, and one of the greatest is that he, too, would now find himself locking horns with the learned consensus. In the extract above, I left out a phrase at the end of the first paragraph. What Frank actually said was 'so as to be able to understand their wonderful structure and mechanism, *and the handiwork of the great Designer of all things*'. The second paragraph concluded: 'It is hardly necessary to say that I am not a disciple of Darwin . . . I believe in the doctrine – I am sorry to say now old-fashioned – that the great Creator made all things in the beginning, and that he made them good.' This does not divide me from him. Unremarkably for a man of my age and times, I believe in the non-existence of God.

But so what? I can still listen to Frank and be mesmerised by his brilliance. Though his declarations of faith become lengthier, more frequent and occasionally tedious, they do not devalue his science. His self-description as 'Zoologist' in the title of *Log Book* is a simple statement of fact. Science and religion share common cause in the pursuit of mystery. Like art, science exists independently of whatever inspires it. God or Darwin, why should it matter?

In 1873 there was better news for salmon. Frank's evidence to a parliamentary committee helped to ensure that the Salmon Fisheries Amendment Act of that year would be a genuine improvement over its near-useless parent of 1861. The original Act had imposed some restrictions on fishing (no netting after August; no angling after October), had laid down annual and weekend (noon Saturday to 6 a.m. Monday) close times, and supposedly made fish passes compulsory at dams and weirs. Cue hollow laughter. The Act was scientifically illiterate: it ignored the way salmon actually behave. 'Since the Salmon Act of 1861 was passed,' Frank observed wrily, 'we have discovered that the salmon will not obey the laws which the legislature enacted . . . Salmon will not ascend rivers according to Act of Parliament, but will come up just when it pleases them to do so . . . The new law takes cognisance of the habits of the fish.' Essentially now there would be a bit more give and take. Netsmen on some rivers would be allowed to work in September, and the rods could come out on certain days in November. That way, said Frank, more salmon would find their way to market, and anglers could have a bit of extra sport. To balance all this, new by-laws could extend the weekly close time, bailiffs would have extra powers against poachers, and the sluices at mill wheels had to be kept shut on Sundays and at all other times when the wheels were idle. 'The consequence of this', Frank explained, 'will be, that on Sundays, throughout the length and breadth of the land, the water . . . will by law be sent over the weir; so that the ascending fish will have an extra chance of getting up, while the descending fish will be enabled to get down to the sea, earlier in the spring than they did before.'

The 1873 Act also banned the harvesting of baby eels, or elvers, millions of which were being taken from the Rivers Severn and Wye in April and May. This caused a bizarre controversy. Magistrates were reluctant to enforce the law because it was argued that elvers were not baby eels at all and thus were fair game. This sent Frank scurrying to Gloucester, where he settled the matter by displaying specimens of eels at different stages in their life cycle and thus removing all doubt about the elvers' parentage. In the same year, more in hope than expectation, he took the young salmon from his hatchery at the museum and released them into the Thames. Cynics predicted that the tiny fish would get no further than the jaws of the nearest pike or perch. Frank agreed they might have a point, but he hoped a few at least might escape. The result was one of the flights of whimsy that his critics so loved to decry. It must be remembered, he said, 'that these are learned fish, and have been reared in the Science and Art Department in South Kensington. They ought therefore to have more science and art in their little brains, than the vulgar common little fish hatched in a gutter.' But he had no illusions. The fact was, he confessed: 'These little fry have not the same amount of intelligence as fry found in a wild state.' Instead of darting away as wild fish would have done, they 'gaped about' just waiting to be eaten. While Frank's enthusiasm for pisciculture never faltered, he realised that making it easier for wild salmon to run upstream to their breeding grounds, and downstream to the sea, was much more important. 'This is what I call real fish culture, carried on over many hundreds of miles of river and mountain-brooks in our own country.' It had been only four years since he had written so excitedly to *The Times* about the salmon at Gravesend. It would be another 137 years before the EU-funded researchers concluded that 'habitat reconstruction' was a better bet than restocking. Just as Frank had said it was.

The winter of 1874 was unusually severe. Frank himself must have shivered in the icy rivers, but his anxiety was for the zoo. Many of the animals were strangers to sub-zero temperatures; only the polar bears, beavers and seals had any cause to be happy. The rhinoceros was barred from its pond until the thaw. Elephants, giraffes and elands were kept warm in their houses by hot-water pipes, and the hippos' bathing water was raised to fifty-five degrees Fahrenheit. Snakes, literally shocked rigid by the cold, were kept at ninety degrees and wrapped in blankets. On top of all this, Frank had another *cause célèbre*, about which he would write to *The Times* in March 1875. Three years earlier, along with Captain David Gray of the sealing and whaling ship *Eclipse*, he had proposed a close season for Arctic seals. Their concern was focused on the island of Jan Mayen. Every year this tiny volcanic isle yielded £250,000 worth of seal oil, and the smell of money had performed its usual perverse magic. Every spring, Frank informed readers of *The Times*, a crowd of sixty ships – mostly Scottish and Norwegian but with others from Germany, Sweden and Holland – would set about their work of destruction. They began killing just as the pups were born in mid March. Unless they were stopped, said Frank, 'the seals will be utterly destroyed'. As every anti-seal-cull protester has done since, he wanted to make clear how cruel it all was. He quoted Captain Gray:

Last year the fleet set to work to kill the seals on March 26 . . . and in forty-eight hours the fishing was completely over, the old seals being shot, wounded, or scared away, while thousands upon thousands of young ones were left crying piteously for their mothers. These mostly perished of famine in the snow, as they were not old enough to make worth while the trouble of killing them. If you could imagine yourself surrounded by four or five hundred thousand babies, all crying at the pitch of their voices, you would have some idea of the piteous noise they make. Their cry is very like that of a human infant.

When a close season was agreed in 1876, Gray had no doubt about whom to thank. After Frank's death, though he faced stiff competition, the good captain would make a serious bid for the title of most fervent admirer. 'I first knew Mr. Buckland personally in 1870 . . . he was the man I wanted most to know in the world.' Among the many others who wanted to know him were the royal family, for whom he had become a kind of de facto Gentleman of Field and Pond. In July 1875 Prince Christian called him to Windsor Castle, which was afflicted by a plague of frogs. The Queen was 'much pleased' by Frank's advice to 'turn out the ducks'.

While all this was going on, Frank kept up his regular stream of chatter in *Land and Water*. Under his tutelage readers learned about the structure of fish scales, crocodile skin, the scaly armour of pangolin, armadillo and tortoise, the spines and claws of the hedgehog, the sonorous intricacies of bell-ringing, the dried heads of Ecuadorian 'Indians', the uses of horsehair. Having measured the size and weight of horse tails, he concluded that the best ones came from hearse horses. 'The largest horse's tail I ever examined weighed two pounds two ounces; the hair was no less than six feet in length. The horse belonged to Mr. Ebbutt, undertaker, of Croydon.'

Frank's enthusiasm for acclimatisation remained as strong as ever. He was thrilled by the Duke of Marlborough's emu paddock at Blenheim Palace, and thrilled even more by the kangaroos. Spooked by Frank's sudden appearance, they amused him by dashing for their shed 'in line at a racing pace, like horses in front of the grand stand at Epsom'. He loved watching a baby in its mother's pouch. 'It was very funny to see the little thing's rat-like head peeping out and looking one way while the mother was hopping the other.' Seeing them move, he realised the kangaroo's massive tail was used not 'as a propeller, but rather as a balance'. As an agent of the Acclimatisation Society he had a keen and particular interest in the tail. When eventually he got his hands on one, he found that it contained seventeen bones resembling very closely those in oxtail soup. Good news indeed! 'These sinews and the gelatinous material found inside the tail bones of most animals – fox excepted – make excellent soup, and I am told that kangaroo-tail soup is considered in Australia as good as, if not better than, ox-tail . . . I am very anxious that kangaroos should be cultivated in English parks. I am sure they would do well, and be very ornamental, as well as forming a new dish for the table.' One of the things that made Frank such a pleasure to read was his cartoonist's eye for a visual pun. The way a kangaroo rests on its tail when sitting upright reminded him of 'the curious old seats one sees in cathedrals, so arranged that the monks could lean against them, but if the monks went to sleep, the seats fell down'.

More serious matters were demanding his attention. In 1875 the Home Secretary, R. A. Cross, asked him to inspect the crab and lobster fisheries in Norfolk. It did not take him long to identify malpractice. The crab fishery around Cromer was being threatened by the wholesale destruction of baby crabs, and of crabs carrying spawn (known as 'berried' crabs). At Cromer itself the fishermen generally returned undersize crabs to the sea, but their neighbours at Sheringham were not so fastidious. A merchant told Frank that 28,000 baby crabs, worth no more than a farthing each, had been sold at Sheringham in a single day, many of them put to no better use than bait for whelks. The Cromer men had their own voluntary restriction, prohibiting the landing of crabs less than four inches across, but this had not halted the decline. Since 1868 the supply of Norfolk crabs to the Billingsgate fish market had virtually dried up, and fishermen were sailing for the richer pickings at Wells, Mablethorpe and Grimsby. To keep what remained of their livelihoods, the Cromer men now wanted their voluntary constraints made law. The Inspector of Fisheries did not need to be asked twice: they were talking his language. Unless fish had time to breed before they were killed, then Britain would be an island in a saltwater desert. Frank submitted his report, and in 1876 the Crab and Lobster Fisheries (Norfolk) Act outlawed the possession or sale of crabs less than four inches, and banned the sale of berried crabs – the first ever statutory regulation of a crab fishery. The Act was replaced in the following year by the Fishery (Oysters, Crabs and Lobsters) Act, which added a quarter of an inch to the minimum size and extended the protection of soft and berried crabs across the whole of England and Wales. With his usual pedagogic zeal, Frank attached to his official report a definitive paper on the natural history of crabs and lobsters, examining everything from their breeding habits to their eyelids. Any Member of Parliament who read it would know as much about crustacea as any man alive. The House was also asked to consider a plea from the fishermen of Hall Sands in Devon. As Frank explained, the surf along that part of the coast tended to be frisky and it could be difficult to get a rope from the shore to the boats. The men had got round the problem by training dogs to swim out with the lines. 'The fishermen think it a great hardship that these dogs should be taxed,' wrote Frank. 'We promised to draw the attention of Her Majesty's Government to the matter, and we have noticed it accordingly here.'

Frank also advised a parliamentary committee on oyster culture, which in 1877 granted to oysters the same protections given to crabs. Also in that crowded year he and Spencer Walpole had to conduct yet another inquiry – this one into the practice of harvesting fish with explosives. No surprise, fish-bombing was duly banned by the Fisheries (Dynamite) Act.

No surprise, either, was Frank's reaction to the arrival at London Zoo of a gorilla. Given his quarrel with Darwin, it was perhaps inevitable that he would devote his 'several interviews' with 'Mr Pongo' to drawing up a catalogue of irreconcilable differences between apes and men. He started with the gorilla's way of walking – like 'a dwarf going on crutches' – then moved on to its lips, hair, ear and brain. Alas for Darwin! Unlike human lips, which were designed for speaking, the gorilla's were the dumb lips of a beast. Humans had hair on their heads, but 'Pongo's hair is not hair in our sense of the word, but simply a kind of fur continuous with the other covering of the body'. The clincher was the gorilla's ear:

If the reader will kindly put his or her hand to the ear; he or she will find a very slight little hard knob on the external edge of the fold of each ear, about a quarter of an inch from its highest part. The presence of this knob, according to Darwin, indicates 'the descent' of you and me, my friends, 'from a hairy quadruped, furnished with a tail and pointed ears, probably arboreal in its habits, and an inhabitant of the Old World'.

I was especially careful to examine the gorilla's ear, and I discovered that he *does not wear a knob on his ear.*

What Frank is talking about here is 'Darwin's tubercle', a peculiarity which some evolutionists see as the vestigial remains of a mammalian ear-fold. His error was primarily statistical. Not all his readers rummaging in their ears would have found what they were looking for. Only 10 per cent of adults have Darwin's tubercles, so their absence from any particular ear, human or gorilla, is hardly a phenomenon worth noting. In fact some scientists now think the tubercle owes nothing to our mammalian ancestry anyway. Who knows? Either way, absence of a bump in the ear is not evidence for or against Darwin, or for or against Genesis. Hand-me-down mammalian ears nevertheless do retain a powerful grip on the human imagination. It can be no coincidence that film-makers use pointed ears as shorthand for evolutionary otherness in science fiction and elvish fantasies. In fairness to Frank we must not forget the date. Natural history's knowledge locker in the 1870s was badly understocked. Whereas, for example, Frank had to wonder whether or not gorillas have pineal glands in their brains, anyone now can look up the answer on a website (yes, they do). This meant that his conclusions sometimes owed as much to inspired guesswork as they did to deductive reasoning. We shouldn't wonder that his leaps of imagination occasionally landed him in the wrong place. We should marvel at how often they didn't.

By August 1877 Frank was back in Scotland. Along with Spencer Walpole and Archibald Young, he had been asked to report on the country's herring industry. Moving briskly, the inspectors held court at Edinburgh, Eyemouth, Anstruther, Montrose, Aberdeen, Peterhead, Fraserburgh, Banff, Buckie, Lossiemouth, Burghead, Inverness, Brora, Helmsdale and Wick, hearing evidence from fishermen, fish-curers and the Scotch White Herring Board. So far, so predictable. What took the trip out of the ordinary was an excursion to the Orkneys, Shetlands and some of the remoter outcrops in the west. For this the Admiralty supplied a 340-ton gunboat, *Jackal*, under the command of a Captain Digby. To modern eyes *Jackal* looks the unlikeliest of warships – a paddle steamer rigged as a two-masted schooner – but Frank was mightily impressed. His thoughts, as ever, drifted heavenwards. 'I find that if the roof were taken off the church I attend, St Mary Magdelene, Munster Square, Regent's Park, and the *Jackal* let bodily down into it, she would exactly fit the main aisle. Her bowsprit, however, would project considerably beyond the east window.' The ship had not enjoyed the most exciting of naval careers: built at Govan in 1845, commissioned at Plymouth in 1846, then serving in the Mediterranean until 1851, when she became a store ship at Ascension Island. Oddly, she had been both paid off and recommissioned in 1859, and since 1864 had been employed on fishery protection duties off Scotland's west coast. Frank reckoned her 'a good, smart, obedient creature', with 'a nose as keen as her African four-footed namesake'. They boarded at Wick on 4 September, and very soon found themselves rolling in the Pentland Firth – 'not a disagreeable motion', thought Frank, though he was apprehensive of the Firth's fearsome reputation and was glad enough to drop anchor at Kirkwall. The tightness of the timetable – court to sit at ten the next morning, *Jackal* to sail at three – left him no time to explore, as he had hoped, the fabled 'Druidical remains' at Maeshowe (actually a Neolithic chambered tomb, now recognised as the best surviving Neolithic structure in north-western Europe), but it inspired him to jot down his theory about Stonehenge. This involved inclined planes of impacted snow, and was as wrong as it was ingenious. Some knitted shawls in a shop then set him thinking about a sample of wool which his friend Abraham Dee Bartlett had been asked to identify. 'He proved that it was wool from the large mastiff of Thibet – another instance of my friend's great sagacity.' Bartlett could have been right. The Tibetan mastiff was a hefty breed used in Tibet, China, Nepal and India to keep big cats away from sheep. It was long-haired and, happily for the wool theory, it regularly moulted.

After clearing North Ronaldsay, the first island they saw (though they did not land on it) was Fair Isle, which made Frank think again about knitting. What could explain the intricate patterns of the island's knitwear? He was persuaded that the skills had been learned from sailors of the *El Gran Grifon*, flagship of the Spanish Armada, which was wrecked there in August 1588. The similarity of Fair Isle to Moorish patterns lends some credence to the idea, but it's just as likely that the origins lay with ancient Vikings or passing traders from the Baltic. No one seems to know. Frank's theory is as good as any, though he might have gone a bit too far in supposing that, as a result of the brief (six-week) Spanish visitation in 1588, nineteenth-century Shetlanders had 'a mixture of Iberian blood in their veins'. Why divert himself with such speculations? Like nature itself, he abhorred a vacuum. Where there were questions, there had to be answers.

Questions accumulated with every sea-mile. What, for instance, were the 'large fish' that arrived each November and killed the seals? 'It is said to be from twenty-five to thirty feet long, but not thick in proportion to its length.' Frank uncharacteristically had no suggestion to make, but it seems likely that these were not fish at all but killer whales, or orcas, which have a ravenous appetite for seals, are the right size (adult males grow to twenty-six feet) and are regularly seen around the Shetlands. Of all the contrasts, then and now, this one seems especially stark. In the twenty-first century, any child with access to a television would be able to identify a killer whale; in 1877 even the most eminent marine scientist of his day could be left wondering. An even bigger question – it deserves to be called a mystery – was the 'boat fever' of St Kilda, which was described to Frank by Captain Digby and had been causing misery to the islanders since the seventeenth century.

This boat fever [wrote Frank] is not a new thing; it has been recorded as being in existence a hundred years ago. I think I am close on the solution of this phenomenon. The fever only occurs on St Kilda when a stranger boat lands, and only when the wind is blowing from the east. The cause I believe to be as follows: St Kilda is such a precipitous island that boats can only land when the wind is blowing from the east. When the wind does blow from the east, the people of the island are seized with a kind of influenza cold; and this influenza is imputed not to the wind, but to the arrival of the boats.

This boat fever, or 'boat cough', also interested the Scottish philanthropist George Seton, who visited the islands in 1876. The illness, he wrote, 'usually begins with a cold sensation, pain and stiffness in the muscles of the jaw, aching in the head and bones, and great lassitude and depression – the ordinary symptoms of catarrh in an aggravated form – and is accompanied by a discharge from the nose, a rapid pulse, and a severe cough, which is particularly harassing during the night'. It is possible that Frank read this before he recorded his own comments in *Notes and Jottings from Animal Life*. It seems clear that the vector was not the wind itself but the visitors who blew in with it. 'The malady', wrote Seton, 'first attacks those persons who have come most closely into contact with the strangers, and then extends itself over the whole community.' But neither Seton nor Frank was able to identify the disease beyond the folkloric 'boat fever' or 'boat cough'. Life on St Kilda was tough. By the time of its abandonment in 1930, there was still no running water, sewerage or electricity. A photograph of fishermen taken in 1886 shows many of them without shoes, a fact that added cold feet to a growing list of possible causes of the fever, which also included 'impurities in the atmosphere' and the peculiar toxicity of the Harris ferry. The St Kildans were, in the language of medical men, 'immunonaive' – so isolated that they could not build up the herd immunity to everyday infections which protected people on the mainland. It was not until 2008 that the *Journal of the Royal College of Physicians of Edinburgh* published a paper proposing a scientifically cogent answer. The author, Peter Stride, a doctor from the Redcliffe Hospital in Queensland, Australia, had worked his way through a list of possible viruses and matched them against the symptoms of boat cough. The culprit, he concluded, was none other than that ubiquitous antagonist of the human throat and nose, rhinovirus – 'the most common worldwide infective viral agents in humans and the most frequent cause of the common cold'. Frank would have been happy with that, I think, and would have been impressed that a physician from the other side of the world had cared enough to look into it. A man after his own heart.

Other medical peculiarities in Scotland were superficially stranger, though easier to account for.

I was told at Lerwick of another disease which was new to me. It is a peculiar disease of the eye, brought about by the presence of herrings. To guess at this disease would be almost impossible, but when the facts are known the reason becomes apparent. The scales of the herring are very thin, and like very thin glass. They fit very loosely into pockets in the skin of the fish. The women, in the operation of cleaning the herrings, handle them very quickly; the scales are rubbed off in flakes like snow, and are rubbed from the hands into the eyes; they then get under the eyelids, and can hardly be seen, being so transparent. The only way to get them out is for the operator to open the eyelids quite wide, and lick out the herring-scales with the tongue.

Frank was struck by Lerwick's 'splendid harbour', so big that in 1605 it could shelter a 94-strong fleet of English warships. The first thing he noticed as he was rowed ashore was the fishermen's way of keeping their boats upright at low tide – by wedging them with whale skulls. This waste-not, want-not philosophy of hard-pressed people tended to encourage practices likely to be offensive to modern sensibilities. The RSPCA had been founded in 1824 (though the 'R' wasn't awarded until 1840) and it was largely due to its agitations that the Cruelty to Animals Act 1835 had outlawed maltreatment of dogs, sheep and goats (cattle were already protected by the Cruel Treatment of Cattle Act 1822), and banned bear-baiting and cock-fighting. But care and sentiment were reserved for these few. While domestic animals could nestle in the protective arm of the law, wild animals had to take their chance. So too, law or no law, did dogs that lived too near the sea. Frank described what was likely to befall them:

All over Scotland I observed that the nets and long lines are frequently buoyed by dogs' skins. They catch Mr Dog, kill him, cut off his head, and turn his skin inside out, hair inside; they tar the outside, then tie up his legs, and put a wooden plug into his neck, and blow him up quite tight by means of a plug in one of his legs. They tie the plug on to the buoy rope, and the dog's tail and hind legs floating on the surface of the water have a very curious appearance. For some reason they don't turn cats into buoys, and pigs are too expensive. The month of May is especially dangerous for dogs, as buoys are then wanted. Mr Dog gets a crack on the head, is turned inside out, blown up and tarred, and in a quarter of an hour is anchored to a net out at sea. I think it would puzzle anybody, even a judge at a dog show, to swear to his dog when blown up without a head, and turned inside out.

Whales excited less sympathy than we might now expect. *Save the Whales* was the first great rallying cry of the conservation movement, but the nineteenth century was hungrier and less given to niceties. Politically at least, the weakness of the modern conservationists' case is their tendency to argue that all conflicts between humans and wildlife should be settled in favour of the wildlife. Extreme and often expensive inconvenience is thus imposed on anyone whose life impinges, for example, on a bat. To the Victorians this would have been insanity. Frank saw fin whales as a pestilence, unwanted plunderers of a vital human resource. Far from wanting to protect them, he saw no reason why 'these rascally herring-poaching Finners' should not be harpooned by gentlemen for sport. Different century: different priorities.

For all his love of Scotland, Frank began to feel he'd had enough of its endless scenery, 'everlasting mountains and solitude', and its droning monotony of names. 'It is always Ben this or Glen that, every day and all day long.' He complained that there was no such thing in Scotland as a hill: every molehill was a mountain. Though he admired Archibald Young's encyclopedic knowledge of his homeland – the height of every

mountain, the depth of every loch, length of every river, name
of every owner of a shooting lodge – there was something in his
tone that maddened him. His general impression, he remarked
sarcastically, was that no one in Scotland had less than £20,000
a year, and that the correct thing for a man to do when he had
made his fortune was 'to find out a desolate, barren island where
Robinson Crusoe himself would be uncomfortable, or a lonely
moor where there is nothing but barren rock and heather'.
His explanation for this carried a sardonic whiff of the
yet-to-be-invented science of evolutionary psychology. Man in
his primitive state, he reasoned, lived like an animal by hunting.
There was no other way he could survive. So, all these millions of
years later, what happened when a gentleman had accumulated
every luxury money could buy, and had more food in his larder
than he could eat? 'Why,' said Frank, 'he immediately goes back
to his primitive state, and begins to hunt again.'

On his earlier tour of the salmon rivers he had been easier
to please. Then he had enjoyed 'meeting many gentlemen
interested in the noble sport of deer-stalking'. The red deer,
he thought, was very clever, 'not nearly so stupid as sheep'.
Literally to give weight to his theory, he had put a stag's brain
on the scales. '[It] weighs one pound one ounce,' he reported.
'A sheep's brain weighs four ounces, and two sheep about equal
a 16-stone stag in weight; therefore a sheep's brain is, in proportion to the size of the animal, only half as large as that of a
deer, and it will require four sheep's brains to equal the brain of
one stag.' Thinking about deers' heads steered him dangerously
close to Darwin, though he didn't see it that way. The deerstalk-
ing gentlemen complained that the animals' heads were getting
smaller. Of course they were, said Frank, and it was the fault
of the gentlemen themselves. He compared the deer with race-
horses. 'Those who breed race-horses select the finest males

for stud, and the breed gradually approaches perfection. Those who breed red deer, on the contrary, select the finest males *for the rifle*; the moment a stag grows horns larger and finer than his brother stags he is doomed to death; everybody is after the "muckle stag of Ben-something or another"; he is ultimately shot, and the breed is not benefited by the continuance of his kind.' What the racehorse breeders were doing was mimicking the process of sexual selection by which females choose the most desirable mates to sire their young, and which is one of the principal engines of evolution. What the deerstalkers were doing was frustrating it. Frank saw cause and effect well enough, but he missed the underlying principle. If he believed, as he insisted, that 'the great Creator made all things in the beginning, and . . . he made them good', then the idea of man-made 'improvement' looks almost blasphemous. Ignoring this, and with scant regard for the Creator's grand design, Frank proposed a typically Bucklandesque solution to the gentlemen's difficulty. The deer should be crossed with wapiti. 'I am sure it would . . . greatly improve the heads of deer in their forests.' But the Scottish love affair seems gradually to have faded through overfamiliarity. In 1877, as August flowed into September, the landscape so grated on Frank's temper that he 'positively refused to look at any more scenery of any kind', and declared his hatred of the word *picturesque*. Somewhat wistfully at Oban he saw an omnibus and a four-wheeled cab, both teasing reminders of home.

It is probably just as well that the forests were never inhabited by wapi-deer. We understand now that human intervention in nature's affairs carries a high risk of unintended consequences. The twentieth century provided plenty of examples. When animal rights activists 'liberated' American mink from British fur farms, the effect was a death sentence for water voles, which were saved only by a counter-balancing and ultimately extreme human intervention – extermination of the mink. Scotland suffered a similar misadventure in 1974 when some animal lovers released hedgehogs on South Uist. By 2000, after an egg-eating orgy that lasted twenty-six years, the original small colony had multiplied to many thousands and had spilled over to Benbecula and North Uist. For ground-nesting birds it was Armageddon. To save them, thousands of hedgehogs had to be either killed or shipped to the mainland. In the 1870s truths like this had yet to dawn. Frank nevertheless recorded what might have been the earliest example of conservation policy backlash. In 1869 the British government had passed the Sea Birds Preservation Act, the country's first ever legislation to protect wild birds. Though he neglected to mention the fact, Frank himself had served on the committee that framed it. Now the gulls were coming home not just to roost but to pillage. Huge mobs of legally protected black-backed gulls were terrorising salmon fisheries and destroying more fish 'than all the other salmon poachers in Scotland'. They were also stealing oats and wrecking root crops. In a six-acre field, wrote Frank, the birds could destroy 'as many turnips as would feed three cows'.

Nor were these the birds' only crimes. The statistics of the herring fishery were almost too astronomical to grasp. A single shoal recorded in 1877 was eighteen fathoms (108 feet) deep, a solid mass of herring stretching four miles in one direction and two miles in the other. Frank calculated that about a million barrels of herring, or 800 million fish, were taken every year by Scottish fishermen, whose nets joined together would stretch nearly 12,000 miles and cover seventy square miles. If one assumed that the English, Irish, French and Dutch fleets all fished at the same rate, then the total annual take would be 2,400 million, which must be reckoned a pretty heavy hit on a single species. And yet, despite all this and to our retrospective astonishment, the inspectors concluded that 'the destruction of herring by man is probably insignificant compared with that wrought by other natural agencies'. It was those pesky birds again. 'Nothing that man has yet done, and nothing that man is likely to do,' they declared, 'has diminished, or is likely to diminish, the general stock of herrings in the sea.' It would therefore be 'inexpedient' to place any restrictions on the right to fish. Indeed, existing restrictions on close seasons and mesh sizes should be repealed, and so in Scotland should the Sea Birds Preservation Act, which prohibited the slaughter of gannets.

After Scotland came England and Wales. In the following year Frank and Spencer Walpole were instructed to extend their inquiry to the rest of Great Britain. It occupied nine months of 1878 and a good part of 1879, when it was extended to the Isle of Man. It would be, in John Upton's words, 'the most exhaustive account of our marine fishery ever compiled'. That much might have been expected. What nobody yet understood was the toll it would take on Frank. It would be the end of him.

CHAPTER FIFTEEN

Snow in Norfolk

One of Frank's unlikelier enthusiasms was for bluebottles. He valued them, as he valued rats, for their effectiveness in cleansing the world of decay, and for their extraordinary sensory abilities. If he could marvel at the strength of an elephant, then he could marvel equally at the bluebottle's uncanny powers of scent. As usual, he made his point with a story:

> A poisoned rat had crawled under the floor of a very smart dining room, and had died there. The room in consequence became uninhabitable for a time. All the boards of the room were about to be removed, when somebody suggested that a blue-bottle fly should be turned out. Mr Blue-bottle hunted about the room for some time; at last he lighted upon a certain spot on the floor; the single board was removed, and there, sure enough, was the dead rat.

The first rule of nature, 'Eat and be eaten', was also nature's first rule of housekeeping, and flies were the perfect example. As the only creatures that could kill an elephant, which they did by infesting its wounds with maggots, they were also more powerful than they looked. You might almost say this was the first rule of Frank. The ubiquitous was always as diverting to

him as the exotic. One day in Windsor Great Park when he was
supposed to be stalking deer, he stalked a mole instead. He then
released it onto a lawn, where red tape attached to its leg enabled
him to track it underground: speed and direction duly noted.
Afterwards he took a train back to London, all the way observing
the mole in a biscuit tin. Back at Albany Street his notes expanded
into a lengthy dissertation composed in the mole's own imagined
voice and quarrelling with a paper in the *Proceedings of the Royal
Society*. The author's mistake was to have suggested that the
mole's poor eyesight was a product of evolution. The mole, as
voiced by Frank, was indignant. It insisted that the 'same great
wisdom' which had designed its nose and fur had also designed
its eyes, precision-made at the moment of Creation for their
life's work of finding worms in the dark. 'I defy any philosopher
in the world to disprove this,' it said.

The philosophers might also have been surprised by Frank's
assertion that the navy had moles and rooks to thank for its milk
supply. The reason was simple: both kept down wireworms, and
the grass was more nutritious after moles had passed beneath.
Frank's source for this intelligence was, untypically, not his own
observation but 'a merchant in gingerbread-nuts' from Harting
in Sussex. None of this meant he had lost interest in bigger
beasts. In November 1879 London Zoo's female tiger caused a
furious fight when her claw by accident tore through her mate's
septum. Frank ended his two-page account of the battle with a
word of advice: 'The best way to stop tigers, cats, dogs, monkeys,
or even men and women fighting is to squirt water strongly into
their faces. The effect is marvellous. Try it.'

He had a less happy time with newts. Their poisonous skin
secretions fascinated him, though dissecting them made him ill.
Dogs evidently felt the same: you never saw one hold a newt
in its jaws. Frank thought the secretions might explain the

salamander's supposed immunity to fire – the exuded moisture might allow it to survive just long enough to crawl from the flames. But not, sadly, long enough to withstand the tortures inflicted by alchemists. Frank was both thrilled and appalled by a 'M. Bonnet' – presumably the eighteenth-century, Swiss-born French naturalist Charles Bonnet – who, with lofty insensibility to his subjects' pain, had discovered all he needed to know about the animal's other fabled (and this time true) ability, to regrow lost limbs.

Frank did not go in for vivisection: his own dissections were confined to the dead, and were never less than thorough. In his posthumous *Notes and Jottings from Animal Life* he recalled a 'huge otter' which had been caught near Yarmouth's North Pier. His investigation of its innards could hardly have been more particular:

> Holding up the pharynx, I poured down thin plaster into the stomach, which, of course, hardened, showing its full capacity; it was nine and a half inches long, and fifteen inches round, and would hold rather over three pints of fluid. The oesophagus . . . strange to say, was a very small tube, the size of a half-inch gas pipe, hardly big enough to admit one's little finger, and only one inch and three-quarters round. I expected to find it a large dilatable tube, as in other fish-eating creatures: why the gullet should be so small I do not know.

He asked his readers to notice the difference between the otter, which chewed before it swallowed, and the seal, which bolted fish whole. Utility was central to Frank's thinking. Why did beavers have flat tails? It was commonly believed – 'not only in school-rooms', Frank complained, 'but much higher' – that they were used like trowels to flatten mud. He blamed this wholly avoidable error on Samuel Griswold Goodrich, an American author who wrote for children as 'Peter Parley'. Incredibly, I find the belief still persists 155 years after Goodrich's death. As Frank rightly observed, the tail works as a rudder when the animal swims, and as a prop when it sits or stands. If a resting kangaroo reminded Frank of a monk in a cathedral, then a beaver was more like a mylodon (a giant sloth, known from fossils, which fed on tree branches while sitting upright). Would any modern writer think of this? Would any reader now have a clue what he meant?

Reintroducing beavers in the twenty-first century is a prime ambition of the 'rewilding' lobby which wants to restock Britain with lost fauna – not just beavers but lynxes, wolves and bears. Perhaps unwisely, I told an audience of farmers that I would eat my trousers if a proposal to release lynxes in Norfolk ever succeeded. If I'm wrong I may renege on the trousers but will gladly eat my words. In 2015 as I write, the Scottish government is debating whether or not to allow beavers to remain in Knapdale Forest, Argyll, where they were imported for a five-year trial. England has a small but healthy colony in the River Otter in Devon. Frank – need I say? – was a long way ahead of them. In 1872 he had set about finding some beavers for the Marquess of Bute, who wanted to establish a breeding colony. After two years of patient letter writing he managed to obtain two pairs, one each from France and America, which were released on the Isle of Bute. It was a disaster. The two pairs fought, the animals escaped and Frank was soon dissecting corpses. He did not pass up the opportunity for research, but he regretted the waste (beavers came at £70 or £80 a pair). The beaver's gullet, he found, is small like the otter's, and the small intestines are 'very long and small; the larger intestines . . . exceedingly capacious. The colon is the size of a quart pot. The whole of this part was gorged with gnawed wood.' And there was another myth for him to dispel. 'The beaver is not a fish-eater, never was, and never will be . . . This mistake about the beavers eating fish has been lying dormant for about six-hundred years, and it is pretty nearly time that the mistake should be rectified.'

In January 1875 the marquess bought eight more American* beavers from the London dealer Jamrach. These were given time to recuperate at the zoo, in 'two beautiful dens close by the wombats, just behind the kangaroo sheds', before travelling on to Bute. This time they were sent with written instructions from Frank and Abraham Dee Bartlett, who by trial and error had hit upon the perfect diet: Indian corn, carrots, biscuits and willow boughs. For accommodation, said Frank, 'a kind of pigsty should be made for them, in which a warm nest of straw should be placed, and that they should then be allowed to gnaw them-selves out'. Events then took a happier turn. One of the original four beavers reappeared having been presumed dead, and the newcomers thrived. When Frank stopped off from HMS *Jackal* in 1877, the colony had expanded to twelve. He observed and made notes.

> From the structure they have made, it is evident that they work with a design, I may even say with a definite plan. The trees have been cut down in such a manner that they shall fall in the position in which the beaver thinks they would

---

* The provenance of the beavers has been disputed, correspondence in the Bute Archives suggesting they were Scandinavian. This may be the result of a misunderstanding. As Frank explained, they did not come directly from North America but had spent three months in Germany en route. Though this does not explain the reference to Scandinavia, it does allow the possibility of confusion. As Frank himself was involved in the supply chain, and as his account was contemporaneous, I am inclined to take his word for it.

be of the greatest service to the general structure, generally right across the stream. The cunning fellows seem to have found out that the lowest dam across the river would receive the greatest pressure of water upon it.

By 1878 the numbers were up to sixteen. But it did not last. The marquess's beaver keeper, Joseph Stuart Black,* died in 1887 and the beavers did not long survive him. By the time Black's wife died in 1898, none were left.

At Rye during his sea-fishing inquiry in 1879, Frank somehow acquired a brown hare. He never managed to tame it, but it set him thinking. Hares, he observed, were of varying ability, 'some being very stupid, others very clever'. The best way to tell the difference was to measure the forehead. 'Experience has laid it down as law that, with hares as with men, the more brains they have in their skulls the better learners they become. This is particularly the case with horses: reader, please observe for yourself.' Science has proved him right up to a point. There is a small correlation between brain size and intelligence in humans, but it is one of many genetic and environmental variables and is not definitive. Size on its own can be misleading. Bigger bodies need bigger brains, but most of the extra weight is cortex (the outer layer of the brain), not grey matter. If it were otherwise, then the world's deepest thinker would be the sperm whale, whose 18lb brain is more than five times heavier than Einstein's. It is all about brain-to-body ratio, not sheer weight or bulk. As his experiments with deer and sheep brains showed, Frank understood this. His comparisons were between hare and hare, horse and horse, not hare and horse or horse and man. The one

* Black wrote a report on the beavers for the *Journal of Forestry and Estate Management*, published in 1880.

mystery he never cracked was why the hare needed hairy feet. He could see no reason for this, beyond the fact that they made 'good hat brushes'.

In 1875 Frank had edited a new edition of Gilbert White's classic of observational nature study, first published in 1789, *Natural History of Selborne.* This drew his attention to birds. Through 1877 and 1878, in *Land and Water* and the *Daily News,* he wrote a series of articles about their migrations – a subject which White himself had pondered a hundred years earlier. The one thing everyone knows about eighteenth-century naturalists is their belief that swallows hibernated in mud at the bottom of ponds. Frank's own passion for observation was wholly in the spirit of White, and just as obsessive. He explained how birds' eggs were shaped and coloured to match the size, structure and siting of their nests – yet more proof (as if any were needed) that 'even in this lightly considered department of natural history it is possible to gather convincing evidence that the works of Nature are not the results of mere chance, but are the outcome of a purposeful intelligence'. He watched nightingales, ducks, swallows, martins, linnets, jackdaws, pipits and blackcaps. Jackdaws endeared themselves to him through the sheer awfulness of their behaviour. He noted their habit of stealing baby sparrows to feed their own young; noted, too, that jackdaws themselves made good eating when 'cooked like young rooks'. To the last, he could not understand how anyone of sound mind could live in ignorance of nature. All migratory birds, he said, could be found within a ten-mile radius of London, so Londoners could have 'no excuse for not knowing the notes of birds'. Frank himself was enthralled by these great seasonal movements in the sky.

> While the inhabitants of this great city are fast asleep, during the dark nights that occur generally about this period of

September, many wonderful events are going on high up in the air, far above our heads. Of the nature and cause of these phenomena the general public are little aware. The noises proceeding from the numerous creatures, that are passing over our towns in mid-air, during the darkness of the night, would in former days have probably been put down to the supernatural agency of ghosts and goblins. Observation, however, has taught us that the mysterious forms, shadows and cries proceed from flocks of migratory birds, passing from one part of the earth's surface to another.

He wanted other people to share his admiration for them. 'Quails are said to accomplish a hundred and fifty miles in a night, and undigested African seeds and plants have been found in the crops of these birds when they reach the French coast.' Ducks could manage 1,500 miles in one go; swallows and martins might travel 900 miles in twenty-four hours.* A wise man should always watch the sky. Migrating birds were reliable indicators of changing weather, a fact which, if properly understood, might have changed the course of history.

If the Emperor Napoleon, when on the road to Moscow with his army in 1811, had condescended to observe the flights of storks and cranes passing over his fated battalions, subsequent events in the politics of Europe might have been very different. These storks and cranes knew of the coming on of a great and terrible winter, the birds hastened towards the south, Napoleon and his army towards the north.

---

* The champion long-distance flyer is the Arctic tern, which achieves annual round trips between the Arctic and Antarctic polar regions of approximately 44,375 miles.

Frank took equal pleasure in watching his fellow humans. Bus tops were favourite vantage points, whence he developed a strange belief in the power of coincidence. 'Will my readers kindly notice, that if they see one thing of a peculiar kind they are certain to see its ditto, if not that day, then very soon afterwards.' He supported the theory with evidence. One day on the corner of Vigo Street he spotted a bandy-legged, severely deformed little boy whose feet seemed to have been grafted on backwards. Soon afterwards he saw a cripple on a board, paddling himself along with sticks. Two in ten minutes! His clincher was the Day of the Red-headed Girls. At Regent (now Oxford) Circus he looked down from the bus and noticed a girl with an immense mop of fiery red hair. He counted the minutes. Another redhead appeared opposite Negretti and Zambra's* in Regent Street; then another near the Life Assurance Office, and another at the corner of Waterloo Place by the Athenaeum. ' "So-ho," I said, "my theories are right, three red-headed girls in seven minutes!"† All the red-headed girls seemed to have come out together the same day.' He scanned the streets for the next four days and did not see a single one.

His observations sometimes led to acts of kindness. On a bitter day of wind and sleet he watched a blind man being led across the street by a dog on a string. He talked to the man, wrote his story and raised a fund of more than £40, enough for man and dog to live 'in comfort instead of privation'. He did the same for his old friend the one-handed fisherman George Butler ('Robinson Crusoe' of Chapter Nine). He published the old

---

* World-famous scientific and optical instrument-makers who had supplied Darwin's voyage on the *Beagle*.

† By his own account there seem actually to have been four of them.

man's address – 3 Willow-place, Mill-lane, Forton, Gosport – and begged readers to send him books and hire his boat. More usefully, perhaps, Frank himself persuaded the Admiralty to raise Crusoe's pension.

His schoolboyish sense of humour was always likely to get the better of him. In Scotland he dropped a parcel of stinking fish in the street so that he could watch the reaction of whoever picked it up. For no better reason than to bamboozle witnesses at an inquiry, he once made 'herring roe' from tapioca and whisky. None of this affected his popularity. Visitors continued to turn up at Albany Street on a bewildering variety of errands. The Chinese ambassador came to learn how to use a fishing rod, the Chief Rabbi to talk about oysters. He wanted to know: *did they creep?* If they did, then according to the laws of Moses Jews could not eat them.* Though unrivalled in his expertise, Frank thought it right to consult the Dean of Westminster, Arthur Stanley, before delivering his verdict. Sadly for Jewish gourmets, oysters did indeed creep. No such inhibition affected the Chinese ambassador, who attended a grand party at Albany Street in November 1878, where, in Bompas's account, 'men of science and art and social rank' mingled with 'dealers in wild animals, bird-fanciers and fishermen'. For their amusement, Frank provided a fine range of oysters chosen to 'illustrate their variety and grace the entertainment'. Alas, with his peacock's feather nodding above the multitude, the ambassador intercepted them on their way upstairs, 'and before anyone could

* The question seems a strange one, as all shellfish are prohibited by the admonition against finless and scaleless fish. Leviticus 11:10: 'And all that have not fins and scales in the seas, and in the rivers, of all that move in the waters, and of any living thing which *is* in the waters, they *shall be* an abomination unto you.'

find words or signs to describe their scientific rarity, devoured them all'.

It was not just the sea-fishing inquiry that was draining Frank's energy. The New Zealand government needed more salmon ova and, though it was far too late in the season, Frank had promised to find them some. The ova would travel from London to Melbourne aboard the steamship *Chimborazo*, which was due to sail on 21 January 1878, leaving Frank precious little time. In the second week of the month he set off by train for Newcastle ('writing all the way, of course'), thence by train to Chollerford, followed next morning by an eleven-mile drive to Bellingham and another train to Reedsmouth, where a team of local men were waiting to help him collect eggs in the North Tyne. They were met by rising water and vile weather.

> With the spate in the river, came the storm upon us, a regular spiteful gentleman fresh from the caves of Eolus,* iced rain, sleet, and snow. I was cold, very cold, but I would not let it be seen. I felt my wet suit of waterproof gradually freezing and becoming like a suit of armour, especially about the arms and throat; so we packed up and walked away as fast as we could, and got a sort of shelter under a railway arch, where I managed, with the help of a water bailiff, to get off the frozen dress; and then for a walk – I hate walking – into Bellingham. As we went along, a blacker cloud came over, and it began to snow, not in nice heavy flakes, but little sharp cutting spikes the size of peppercorns. The howling wind drove them along like a volley from an infantry regiment.

* The reference is to the Greek god Aeolus, king of the winds.

This was not good for him. As he had advised the New Zealanders, it was three weeks too late to be certain of finding ova. But he had promised to try, and try he would, no matter what it cost him. So he packed up, waited several hours at a country railway station and by stages made his way to Carlisle and the River Caldew. It was madness. In he went, up to his armpits, in one vain attempt after another. The spawning season in the Caldew was finished. But there was one last chance: the Devonshire Avon, where salmon spawned later than in the northern rivers. This time Frank did not go himself but sent a man in his stead. A few days later the man reappeared at Albany Street, 'with a face radiant with joy'. To Frank's great relief, he had caught a netful of ripe fish from which 'the eggs ran out . . . like shot out of a shot belt'. The ova were boxed and carried to West India Docks, where Frank himself stowed them in the ice house with some more from his old friend James Youl. This was harder work than it sounds. *Ten tons* of ice went on top of the eighteen boxes of eggs.* 'When we first got into the icehouse,' wrote Frank, 'it was jolly cold, but we were so busy shifting the blocks of ice that we soon had no time to be cold.'

Frank would never collect salmon ova again. Bompas saw the North Tyne as a turning point. 'The long working in icy water, the clothes stiffened with frost and chilled with driving sleet, undermined even his strong constitution, and laid the seeds of disease which soon afterwards developed.' A year later Frank was back at the docks, in the ice house of the SS *Durham*, packing more boxes for Australia. This time the collecting had been done by others. Even so, it was more than his body could stand. 'A few days later,' wrote Bompas, 'he was attacked with inflammation of the lungs, followed by haemorrhage.' He did not leave his room

---

* By Frank's calculation they contained between them more than 20,000 eggs.

for ten days, and it was two months before he could resume his official duties. Yet he was never idle. He treated the illness like a train journey, writing articles and finishing his annual report. This in itself was the culmination of three and a half months' labour, involving a detailed account of every salmon river in England and Wales.

His return to public life began with a stumble. On 23 April he took the chair at a Society of Arts lecture but felt so ill that he had to withdraw. By June, however, his life had resumed its usual headlong pace. Foreign visitors continued to descend. A German representative came to discuss a fishery exhibition scheduled for Berlin in the following year (as ever, there would be no contribution from the British government). In July, six young Zulus arrived in London. Frank studied their appearance (like the 'statues of black marble, or bronze figures one sees in the Paris shops'), their physiognomy ('by no means disagreeable; I could find many much worse faces in the slums of London'), their dances ('emblematic of fighting, and victory to the death over their enemies') and their fearful prowess with the assegai (iron-tipped spear or javelin), which they demonstrated on wooden targets. 'If the object aimed at had been an ordinary man,' Frank observed, 'every assegai would have penetrated his chest.' What tickled him most was an incident at the zoo. 'A good-looking young lady at the refreshment department brought the chief some iced water. [He] immediately wanted to buy her, and with seriousness, asked how many cows her father would take for her.'

Frank's workload over the next few months would have broken the back of many a younger man in perfect health. The sea-fishing inquiry took him to the northern, eastern and western extremities of England and Wales. On top of it, Frank, with Spencer Walpole and Archibald Young, had to conduct another urgent inquiry into a fungal disease which was killing

salmon by the thousand, and which involved hearings at fifteen
different locations across England and Scotland. At one of these,
Archibald Young noted, Frank was obviously unwell.

The robust frame had shrunk and the healthy cheek paled.
He seemed but the shadow of his former self. Yet his interest in his favourite pursuits was unabated and his mind was
as actively vivacious as ever. But sitting in closed rooms and
examining witnesses during a long day exhausted him,
and brought on distressing attacks of asthma.

If Frank had known he was entering the last few months of his
life, then I believe he would have chosen to spend them as he did,
by the sea. He loved the storm-lashed waves, the 'great white
horses' careering up the beach at Yarmouth 'with a Balaclava-like
charge', the dense, gale-driven sand clouds burying the streets.
Even in chaos he could see a pattern. Violence at sea, the 'boiling of the great waters', was 'a happy provision of the Creator'.
Without it, stagnant water 'would become decomposed, lose its
oxygen, and be unfit for the sustenance of fish and the myriad
forms of animal life'. He was a great believer in the healing
power of the sea, and of Yarmouth in particular. So healthy was
it, he said, that the local gravedigger claimed he had been driven
to bankruptcy. But Frank never forgot the sea's unquenchable
thirst for souls. From Yarmouth Pier he watched a crowd gathering on the beach, and was distressed to see the washed-up body
of a boy. 'I was so sorry for the poor lad, and thought of his father
and mother when they heard the news.' He might have thought
also of his own little son and the anguish his death must have
caused Hannah. Had Physie lived, he would by now have been a
young man of twenty-four, the age Frank was when he first met
Hannah in 1850.

Frank learned a great deal about North Sea fishing from a Captain Hill of Grimsby. Hill told him that some eight hundred English boats worked the North Sea and that the trawling ground, stretching right across to Norway, was so vast that the boats were rarely in sight of each other. A reasonable catch for a trawler would be fourteen to fifteen tons of fish, though in fair weather it could be twice as much. In *Land and Water* Frank raised a question that would come back to haunt him. He argued that sea fish, just like salmon, should have a close season while they spawned. Captain Hill agreed, and went further. 'The boats that fish along the shore,' he told Frank, 'should come to sea . . . and fish fair. I should not be surprised if the brood was very soon spoiled and destroyed by small-meshed nets within the three-mile limit.' In 1869 Frank had advised the British Association that deep-sea fish stocks needed careful management. In 1875 he had argued that 'the subject of deep-sea fishing . . . requires further consideration from Government, especially as regards the mesh of nets, and close times within the three mile limit'. He now agreed with Captain Hill. 'It is in these comparatively shallow waters that small-meshed trawl nets are used in the summer months, and doubtless destroy the young [flatfish] fry in tons.' As it was with flatfish in summer, so it was with cod in wintertime.

The cod are killed by a blow on the head, and I was told that the deck of the boat, when cod are being killed for the markets, is *actually milk-white with spawn*, and this fearful destruction has been going on now for years. If I understand aright, the cod are in the height of spawning about the middle of February, and this is when the above lamentable sight can be seen. Surely this must in time produce some diminution in the number of fish, and I wish I could see my way clearly as to advising legislation for the future.

The cod-fishers are, I believe, willing that this fearful
destruction should be stopped.

He could have had no idea at the time how this would reflect
on his reputation. Indeed he could have had little certainty
about anything. While there was no shortage of opinion, there
was a perilous dearth of facts. Yet again Frank appealed for a war
on ignorance. Spawning seasons should be accurately recorded,
he said, so that close seasons could be fixed to protect the fry.
None of this harmed his pleasure in watching the fisherfolk
at work. The floating boxes of live cod at Grimsby – so many
that 'men and boys are frequently seen walking about on top of
them, and jumping from one to the other, over a great extent
of the dock' – was a sight worth going all the way to Grimsby
to see. For an even bigger treat, he advised his readers to take
the Great Eastern Railway to Yarmouth in September, when a
vast fleet queued for the fish quay and buried it in baskets of
herring – so many, thought Frank, 'that it seems a wonder to me
how ever they were going to be disposed of'. It was the same at
Lowestoft – such an immensity of supply that it couldn't be
imagined it could ever run out. Even Frank struggled with the
scale of it. 'The quantity of trawl fish on the quay was something
amazing, and how it is that the North Sea is not trawled out
may be a wonder to many.' But he hastened to reassure people.
The whole of England and Scotland, he said, could be sunk in
the North Sea and there would still be room to sail around them.
It was *that* big.

The inspectors' official *Report on the Sea Fisheries of England
and Wales*, the summit of Frank's achievements, was published
in September 1879. It banged the same old drum. How could
the inspectors be expected to find certainties in such a void of
factlessness? 'The value of the fisheries in this country may be

computed in millions; the capital invested in them in millions; the persons dependent on them in hundreds of thousands; and yet there is no really accurate statistical information upon the subject, and there are no means whatever of comparing by figures their yield now with their yield in former years.' In other words, there was no way of knowing what impact the trawlers were having on the stocks. Despite this, and despite all Frank's fears about the loss of spawn, the report not only reaffirmed the message from Scotland – that there was no risk from over-exploitation of herring – but extended this comforting reassurance to every other species in the sea. Looking back from the twenty-first century, it now seems absurd. It seemed absurd when G. H. O. Burgess published his biography in 1967. Dr Burgess himself was a fish scientist, and an eminent one at that, the man in charge of the then Ministry of Technology's Humber Laboratory for Fish Technology. All the same, I think his verdict on the report as 'pseudo-science' was too harsh. Look at it from Frank's point of view. No one could have fought harder than he did for investment in research. It was hardly his fault that he and Walpole had to rely so heavily on anecdote. They had no crystal ball.

Concern about the effects of human activity on the environment had yet to be widely heard. All the same, the doomsaying had begun and Frank was likely to have known about it. Fifteen years earlier, in 1864, the American diplomat and philologist George Perkins Marsh published his great work, *Man and Nature; or Physical Geography as Modified by Human Action*, in which he deplored 'man's ignorant disregard of the laws of nature'. The slaughter of insect-eating birds, for example, was having dire effects on crops and wild plants. Man, he said, 'is everywhere a disturbing agent. Wherever he plants his foot, the harmonies of nature are turned to discords.' If men didn't change their ways, he warned, then they would reduce the earth 'to such a condition of impoverished productiveness, of shattered surface, of climatic excess, as to threaten the depravation, barbarism, and perhaps even extinction of the species'. In energy and outlook, Marsh and Frank were very similar. Marsh like Frank was a champion of the Old Testament, though he seems to have had a better understanding of Darwin. Like Frank, too, he was more than just a capable observer of natural life. He was a unique interpreter of it, an original thinker who first gave weight to the idea that humankind could degrade the work of the Almighty. Frank was, in his several ways, brilliant. Marsh was a kind of genius, deservedly regarded as the first true environmentalist. But even he could see no imminent danger to fish. 'The inhabitants of the waters', he wrote, 'seem comparatively secure from human pursuit or interference by the inaccessibility of their retreats, and by our ignorance of their habits – a natural result of the difficulty of observing the ways of creatures living in a medium in which we cannot exist.' Here, then, is an important caveat to Burgess's dismay. No one writing in Frank's lifetime could have predicted the destructive force of the worldwide fishing industry in the next century. Not even Burgess in 1967 had seen the worst of it. Exhibit A in any record of oceanic abuse would be the Canadian cod fishery on the Grand Banks of Newfoundland. In 1497, when John Cabot first arrived there, the Atlantic seemed more fish than water. The shoals were so dense you couldn't row a boat through them. By the mid twentieth century giant factory trawlers were flocking there from across the globe – Britain, Germany, Spain, France, Portugal, Russia, even China and Japan steamed in with their billowing continent of nets. From the shore at night, the lit-up fleet looked like *Megalopolis*. Between 1960 and 1975, it stripped the Grand Banks of eight million tons of fish, peaking in 1968 at 800,000 tons in a single year: big

fish, small fish, spawning females, the lot. There was only one
way it could end. In 1992 the fishery collapsed. You could have
drained the sea and not found enough to stock a fish-and-chip
shop. Other disasters had been piling up. Collapse of North Sea
herring in 1978 and mackerel in 1985. Catastrophic declines in
cod, sole, skate and bass. Seabirds in Scotland crashing because
they had nothing left to eat. On top of all this came the ding-
bat politics of Europe's Common Fisheries Policy with its crazy
'bycatches', 'discards' and two fingers to the science. None of
this lunacy could Frank have predicted. All of it he would have
fought against.

Burgess did have a point, though. Frank by now was far from
well. He was a sick man, moreover, whose arithmetic even at
the best of times was shaky. Burgess found himself wondering
whether Frank 'understood the full implications of what was
being said under his name'. For years Frank had campaigned
for legal controls. Yet now he seemed to believe no action was
needed. So stark was the contrast that Burgess felt 'tolerably
sure' that Frank could not have written the 'offending' passages
himself. If he was right, then the author of the deception – one
must assume it was Spencer Walpole – was guilty of, at best,
colossal stupidity, and at worst a reckless act of sabotage. Burgess
was certainly right about one thing. Frank's celebrity by now
rode so high that anything attributed to him would be accepted
without question, *because Buckland says so*. '[The] prestige of
Buckland and Walpole', Burgess thought, 'was such that their
views were quoted as providing the last word on the subject for
the next twenty years, by which time there was no doubt that
they were wrong.' Was Frank's name taken in vain? Burgess
clearly believed it was, but I am not so sure. There is no evidence
from anything else Frank did at the time to suggest he was losing
his grip. And why would Walpole want to wreck the inquiry?

Near the end: a plainly exhausted Frank photographed in 1879. He had
only a year left to live

The inspectors' report is now a rare document, not easy to
find. After a long search, and for an outlay of fifty pounds at a
bookshop in Kent, I found myself in possession of a cloth-bound
volume, 282 pages, octavo, indexed, with fold-out map and
charts. It was printed in 1879 by George E. Eyre and William
Spottiswoode, 'Printers to the Queen's Most Excellent Majesty,
for Her Majesty's Stationery Office', price 1s 10d (9p). An
inscription shows that its first owner was one G. R. Stagg, who
bought it on 1 January 1880, in Great Yarmouth. A pencilled
note at the foot of the page identifies it as 'an important report by
Buckland'. Even today, with all the advantages of technology, a
work on this scale would be an outstanding achievement. For two
men labouring in the age of the dipping pen, it was astonishing.

Frank and Spencer Walpole began hearing evidence at Mevagissey, Cornwall, on 26 March 1878, and finished at Castletown, Isle of Man, on 6 August 1879. Down the length of the east coast, day-long sittings were held at Berwick-upon-Tweed, Craster, Newbiggin, Cullercoats, North Shields, Sunderland, Hartlepool, Staithes, Whitby, Scarborough, Filey, Hull, Great Grimsby (twice), Boston, King's Lynn, Great Yarmouth, Lowestoft and Aldeburgh. In the Thames Estuary at Southend, Leigh, Gravesend, Rochester and Queenborough; on the south coast at Dover, Rye, Brighton, Teignmouth, Brixham, Plymouth, Polruan, Mevagissey, Falmouth and Penzance; on the west coast at St Ives, Instow, Swansea, Carmarthen, Aberystwyth, Liverpool, Southport, Fleetwood, Morecambe, Ulverstone and Furness; on the Isle of Man at Douglas, Ramsey, Peel and Castletown; plus one sitting each at Birmingham and London, fifty in all. Frank's breakdown in health necessitated a six-month hiatus between Hartlepool on 12 December 1878, and Staithes on 4 June 1879, but even so the report was on sale by early autumn. Frank missed only North Shields on 10 July, and the four sittings on the Isle of man in August, which were heard by Spencer Walpole alone. Otherwise he presided over them all.

The impression I get when turning the pages is that Frank's fingerprints are all over them. It is classic Buckland. His notes on the nature of each fish and shellfish – indisputably his own work – run to sixty-eight pages of squintingly small type. Then come thirty-five pages of 'Observations on Certain Special Points Connected with the Economy of the Sea Fisheries of England and Wales', which include a pull-out table on what each species eats – most useful for fishermen thinking about bait. There are notes on the ocean bed ('the great fish farm of the North Sea'), on the growth of sea fish ('little is known'), on migrations and spawning, on the 'structure and peculiarities of flat fish', a table showing the times of year when fish are best to eat, and a fold-out chart showing the numbers and types of fishing vessels at each port, and how many people were employed on them. The numerous references to 'my fish museum', and extracts from private correspondence, never mind Frank's appended signature, leave no room for doubt about authorship. And all this, let us remember, came in an *appendix* to the report proper – a gift Frank need not have made. We need look for no clearer sign of his commitment. 'An important report by Buckland' is a plain statement of the truth.

The summaries of each day's evidence fill 176 pages of the same microscopic type. Witnesses included fishermen, fish merchants and fishmongers, clergymen, naval officers, master mariners, shipowners, lawyers, coastguards, water and oyster bailiffs, conservators, farmers, councillors, constables, town clerks, harbour masters, market inspectors, market superintendents, innkeepers, newspaper editors, medical officers, naturalists, harbour commissioners, net-makers, ferrymen, pilots, Members of Parliament and lifeboatmen. Anyone with a scrap of knowledge or an opinion was given a hearing, and their evidence recorded.

How, then, should we explain Frank's apparent volte face? He had done what he always did, and listened to the people who had most to tell him. The best available experts, the only ones who could see what was actually happening at sea, were the fishermen. In many ways it prefigured the debate that would begin at the end of the next century on global warming. Was a reduction in the catch just an annual fluctuation, nothing to worry about, or was it an ongoing trend? Was it the result of over-exploitation, or could it be explained by natural causes? It was like climate change in another way too. How was it possible that frail humanity, humble servants of the Lord, could pervert or perturb something as mighty, as *infinite*, as the atmosphere or the sea?

Old tensions soon surfaced. Drifters and trawlers blamed each other for damaging each other's nets, and blamed seine-netters and shrimpers for killing spawn. It took time for any consensus to emerge. A smack owner at Hull testified: 'The supply of fish is double what it was 20 or 30 years ago. Trawl fishing does not interfere with the spawn and the fry . . . The trawl cultivates the sea just as the harrow does the land.'

A fisherman with sixty years' experience at Mevagissey: 'There are as many fish in the sea as ever.'

Fishermen at Falmouth: 'No fish are as plentiful as they used to be . . . All fish on this coast are falling off.'

Penzance: 'The present season has been a splendid season for trawlers.'

Great Yarmouth: 'Each smack only catches one third as many [soles] now as it would have caught 10 years ago. This is all over the North Sea. The trade gets worse every year . . . Plaice, turbot, and haddock are falling off in the same proportion as soles.' Yet the Yarmouth men saw no cause for alarm. 'Twenty years back there was a similar scarcity . . . The fish recovered afterwards, and in 1876 there was extraordinary fishing. The failure is due to natural causes, not over fishing, and there is no reason to suppose that there will not be good fishing again.' The fifty smack owners packed into Yarmouth Sailors' Home all agreed that no new legislation was called for.

And so it went on. At Teignmouth, fish were 'as plentiful as ever'. At Brixham, 'there are as many fish caught now as there were 32 years ago. There is nothing against fishing here except the weather.' At Cullercoats: 'It is impossible to exhaust the ocean by taking mature fish.' North Shields: 'The fish in the North Sea are decreasing, and . . . the decrease is due to the destruction of spawn by the trawl.'

Various theories were advanced on where fish spawned, but none could be precise about something they had never seen. If George Perkins Marsh had needed vindication, then here it was. Even those who met fish in their own environment had little understanding of how they lived. This was why Frank had argued so fiercely for research. What he and Spencer Walpole were wrestling with was not 'pseudo-science' but no science at all.

As they saw it, the big question was not whether fishing caused the destruction of fry – it obviously did – but whether the destruction was *wasteful*. On the evidence, they had no option but to conclude that it was not. This was not as perverse as it may sound. All animals bred faster than their capacity to survive, and this was especially true of fish. There was huge overproduction – more than enough, the inspectors reasoned, to absorb the losses without reducing the numbers reaching maturity. Conclusion: 'Speaking generally, there is no reason to suppose that the operations of man are making any sensible impression on the number of fry . . . since there is no evidence that the stock of fish in the sea generally is decreasing.' Where local reductions had been reported, there was 'no evidence that the decrease is due to wasteful fishing or over-fishing'.

And so that was that. No action needed. Three factors convince me that Frank was the principal author. The report bears all the hallmarks of his writing, the conclusions are consistent with the earlier report on Scottish herring, and he repeated them in the revised edition of *Natural History of British Fishes*, upon which he worked until his death. In this, however, he did concede that trawlers caused 'terrible destruction' of young soles, and confessed what he clearly felt was a weakness in the official findings.

> Why, then, are not immediate steps taken to prevent the destruction of young fry? This is one of the most difficult problems my colleague and myself have had to face . . . The question of inventing a mesh of net that will let go the small soles, while it will retain the larger ones (and in my opinion no sole should be taken under 7in. in length) is the greatest problem that we have to consider in dealing with the future Sea Fishery question. This problem, I am sorry to say, is as yet far from being solved; whoever will solve it will be one of the greatest of benefactors to the English people, and would most assuredly deserve a reward not only from the smack owners, fish merchants, and trawlers, but from the public in general.

Were they right to conclude that no legislation was needed? History shows clearly that they were not. Were Frank and Spencer Walpole guilty of negligence, or of misrepresenting the evidence? Just as clearly they were not. It would be a grotesque injustice to blacken Frank's reputation as a naturalist and public servant with the stigma of a blameless failure. In most important ways he has been vindicated. He made clear the desperate need for research, and over time it has happened. He demanded better statistics and these, too, are now available as a matter of course. The government every year publishes numbers by the ream, port by port, species by species, virtually fin by fin. What once had to be guessed at is now in plain sight, and that is Frank's legacy. Now we *know*.

One more thing convinces me that Frank was in good mental shape. Some might conclude that his determination to flog every last drop of energy from his ailing body was evidence of a man no longer capable of judgement. I'd say it showed the opposite: the iron will of a man committed to a cause, whose mind would shine like a diamond until his breath ran out. Bompas had no doubt. 'Although failing health restrained him from bodily exertion and warned him too plainly that his course was nearly run, his intellectual activity did not flag; it seemed rather stimulated to increased exertion, while life remained.' At Grimsby he set up a committee to teach fishermen how to record what they saw, offering prizes for the best reports. Back in London he threw himself into the task of arranging the specimens in his museum – now acknowledged to be superior even to those in the Smithsonian – which he intended to bequeath to the nation. But what an ungrateful nation it was! It would fall to Baron von Bunsen,* vice president of the German Fishery Society, to represent Britain at the International Fishery Exhibition at Berlin in April 1880, where he pointedly contrasted Frank's unstinting efforts with the blinkered lethargy of the British government. 'As I read [Bunsen's] words,' wrote Frank, 'I was overwhelmed with shame.' His exhibits in Germany earned him another gold medal and a diploma, but he would not live to receive them. All the time he continued to work on *Natural History of British Fishes* and articles for *Land and Water.* His word on any issue touching upon fish was the gold standard. In November 1879 he was invited to Billingsgate to give an opinion on frozen salmon from Canada. When properly thawed, he concluded, the fish were 'excellent for the table'. But the month ended less happily. After holding what would turn out to be his last ever inquiry, at Cromer in Norfolk, he was caught in a snowstorm and sank back into illness.

* Son of the former Prussian Ambassador.

It did not stop him writing a chatty article about Christmas at the zoo, but the humour was laced with regret. 'Christmas week for many years past has been to me a period of great importance; for this is the week in which the salmon are at the very height of spawning . . . This Christmas, I regret to say, I shall not have the opportunity of spending my time up to my neck in water.' On 31 March 1880, he presented his final report as Inspector of Fisheries. Consciously or not, it was a last testament, a treatise on all he had learned about salmon. Shortly afterwards he developed dropsy. Reading across the centuries it is not easy to know quite what this meant. Dropsy is a bygone term for what physicians now call oedema, by which they mean a swelling of soft tissue due to retention of fluid. It can occur in various parts of the body but is most common in the legs and ankles. Possible underlying causes include heart or kidney failure. In June, Frank underwent the first of several operations, during which he refused chloroform 'as I wished to be present at the operation'. To rebuild his strength he took a holiday in Margate, where fishermen clustered around him on the beach. He mused upon their likely racial origin – Saxons, he thought – and wondered why it was that men and women were not taught to swim. 'A swimming-master to attend the bathers, at twopence a lesson, would do well here. The bathers could be coached from a boat.' During the sea-fishing inquiry there had been much talk about shrimping, and Margate was the industry's heartland. 'The whole place seems to be shrimps,' he noted, 'one penny a pint.' He reflected on the animosity shown to shrimpers by the deep-sea trawlers, who accused them of destroying fry. Though Frank had downplayed this, he admitted that the trawlers were right. 'If the lower end of the mesh of a shrimp-net be examined, it will be found so small as to be capable of catching a needle.' But he was minded to forgive them. 'Shrimping, it must be remembered, is an industry of considerable national importance, especially as affording occupation for the poor fishermen, food for the poor people of London and the midland counties, and training for men and boys who afterwards join the navy and the merchant service.' The whole problem would be overcome, he said, if shrimpers everywhere copied the men at Leigh, who riddled their catch at sea. Frank's writing at this time displayed the full breadth of his intellectual horizons. As John Upton observed in *Three Great Naturalists*, 'None of his articles are more full of quaint, out-of-the-way information than are those which he wrote in these days of forced inaction.' Other subjects to divert him included shipping in the Channel, coastal erosion and Margate sunsets, the beauty of which he attributed to the effects of sunlight on smoke drifting from London. With unconscious irony and almost poetic symbolism, he noted also that there were 'two seaside places in England where the sun can be seen at this time of the year to rise from the sea and set in the sea'. One was Margate, the other was Cromer.

Even now, 'holiday' to Frank did not mean idleness. Twice he visited the jetty and tried his hand at fishing. Though his catch amounted to one tiny green shore crab, the time was not wasted. He found the spot where the fishermen cleaned their catch, and from the stinking pile of offal was able to dig out specimens for his museum. When he heard there was a sick lioness in the menagerie at the Hall-by-the-Sea pleasure gardens (precursor of the Dreamland amusement park), he was there in an instant. According to Bompas he also bought a capuchin monkey. In August he made what would be his final visit to the Fishery Office. The same month saw the arrival in London of a large orang-utan, which Frank described in the *Daily News*. We cannot know whether or not he was aware his visit to 'Mr Orang' was the last time he would leave Albany Street, and

it is hard to imagine a more appropriate final journey – especially
as it gave him one last opportunity to remark upon the unbridge-
able distance (Mr Darwin, please note) between men and apes.
If his body was failing him, then his mind was forever alert. On
1 September he could even find something to celebrate.

> Her Majesty's Ministers, when sitting down to-day to
> their whitebait dinner at Greenwich, should not fail to rec-
> ollect that the present year commemorates the hundredth
> year of the eating of whitebait. It was in 1780 that one
> Richard Cannon, a fisherman of Blackwall, first introduced
> whitebait as a savoury dish.

It was almost, but not quite, the end. Frank suffered another
relapse and knew he was going to die. His last few weeks and
days were like those of a runner accelerating towards the finishing line. There was so much he had to do. 'Those who saw
him during his long illness', wrote Bompas, 'will well remember
how soon the distressed look of disease would give way to his
old bright smile, and even merry laugh.' The deathbed became a
cockpit of unremitting labour. Articles streamed forth for *Land
and Water*, many of them not published until after he died. For
Macmillan he produced a further revised and cheaper edition
of Gilbert White's *Natural History of Selborne*, including
147 pages of his own notes. He revised and put in order the
articles that would become, posthumously, *Notes and Jottings
from Animal Life*. He judged the Grimsby fishermen's natural
history logs, and finished his new edition of *Natural History of
British Fishes*, whose preface he completed two days before he
died. 'It was thought out in nights of pain,' Bompas said, 'and
dictated in the morning in intervals of gasping breath.' In it,
for one last time, and with the defiance of a man certain of his

destiny, he set out his quarrel with Darwin. 'To put matters very straight, I steadfastly believe that the Great Creator, as indeed we are directly told, made all things perfect and "Very Good" from the beginning; perfect and very good every created thing is now found to be, and will so continue to the end of time.' The last words belong to his brother-in-law, George Bompas:

He likened himself to Job, and wondered why he was so sorely tried; but when the end drew near, he bowed himself to the Supreme Will, and, not unconscious of error, or that a nature so vehement was not always subject to due control, he received the last rites of Christianity, and prepared to die at peace with God and man. 'God is so good,' he said, 'so very good to the little fishes, I do not believe He would let their inspector suffer shipwreck at last.

'I am going a long journey where I think I shall see a great many curious animals. This journey I must go alone.'

Frank Buckland died, aged fifty-four years and two days, on 19 December 1880, and was buried at Brompton Cemetery on Christmas Eve.

CHAPTER SIXTEEN

Life After Death

Frank's obituary appeared in *The Times* on Monday, 20 December 1880. It would have done justice to an emperor. The loss, it said, was not just Britain's. The peoples of Russia, Germany, France and America, as well as Britain's world-wide colonies, were among the many who had reason to mourn. 'No one', it went on, 'has done more to popularize the subject of fishery consultation and preservation, not only in England, but throughout the world.' The obituarist did his best to sum up Frank's lifetime of achievement, but . . .

> it would be impossible to give a full list of his contributions towards the development and preservation of our fisheries, salt water and fresh. His fishery museum at South Kensington . . . is a monument of the unflagging industry with which he sought by all means to gather facts and information in connexion with the fisheries for the public benefit.
>
> As one of the most charming of popular writers on natural history Mr Buckland has endeared himself to thousands who never saw him.

Archibald Young's more personal tribute – 'the most true and genuine man we ever met' – appeared in the *Scotsman* three days later. *The Times* meanwhile embarked upon a strange comedy of errors over the funeral. On 22 December, it announced that the cortège would leave 37 Albany Street at 10 a.m. on 24 December. Next day it altered the timing to 11. In its account of the event, published on Christmas Day, it reported that the coffin and mourners had actually set off 'soon after 1 o' clock'. For all its unpunctuality, the funeral was an affair of considerable size and pomp. The first, open mourning coach was followed by six carriages and pairs, and a cavalcade of private carriages. Hannah rode in the first carriage with George Bompas, whom we must hope was able to get her name right, Frank's sister Mary (Mrs Bompas) and the Reverend H. D. Gordon. Mrs Gordon followed with other family members in the second carriage. Behind them came distinguished representatives from every corner of Frank's richly complex life: men of science, men of the Church, medical and army men, and Prince Christian's equerry representing Buckingham Palace.

In March 1881 Frank was awarded a posthumous gold medal from the International Fishery Exhibition at Berlin. In the same month, the committee of the newly formed Frank Buckland Memorial Fund, chaired by the Duke of Beaufort, met to decide how Frank should be commemorated. The thirty-one-strong committee included two dukes, three marquesses, six Members of Parliament, two professors, three military brass hats (a lieutenant general, a colonel and a lieutenant colonel), the Dean of Christchurch, Spencer Walpole and Archibald Young. Frank's own hero John Hunter had lain dead for sixty-six years before any thought was given to his memorial. Frank himself had to wait less than three months. The recommendations in both cases were strikingly similar. Hunter got a marble statue and a scholarship in his name for comparative anatomy. Frank was to have a bust made in Rome by the English sculptor John Warrington Wood. Hunter's statue stood in the entrance hall of the Royal College of Surgeons; Frank's was to be placed in the institution he had bequeathed to the nation, the Museum of Economic Fish Culture at South Kensington. The committee invited public subscriptions for the memorial, which would also include an annuity for Hannah. Any funds left over would be used 'in some way not yet determined' to promote fishermen's welfare.

All this, one may suppose, Frank would have found intensely gratifying. What might have delighted him even more was the opening at Norwich in the following month of Britain's first ever National Fisheries Exhibition, inaugurated by the Prince of Wales (the future Edward VII). A few days later came a slightly shabbier accolade when auctioneers sneaked his name into an advertisement for an oyster fishery. It was one of many times his reputation would be subject to exploitation. Perhaps the most flagrant was an advertisement for Naldire's Dog Soap, which ran in the classified columns through 1885 and rested entirely on his alleged endorsement: *Naldire's Soap is harmless to dogs, but fatal to fleas.*

On 27 February 1882, at Willis's Rooms in St James's, a beaming Prince of Wales presided over a glittering and eager multitude. Among those gathered were the Duke of Edinburgh, the Duke of Teck, the Duke of Richmond, the Netherlands minister, the ambassadors of Germany, Turkey, Russia, Austria and Italy, the Swedish, Spanish and Brazilian ministers, the French Consul General, the Duke of St Albans, the Duke of Sutherland, Earl Granville, the Marquess of Salisbury, the Earl of Kimberley, the Marquess of Londonderry, Lord Tenterden, the Earl of Melville and Leven, Lord Suffield, the Marquess of Hamilton, Admiral Sir Henry Keppel, Major General Lord Abinger, the Lord Advocate of Scotland, the High Commissioner for Canada, the

Lord Mayor of London, the Lord Provost of Edinburgh, the mayors of some thirty provincial towns, the Prime Warden of the Fishmongers Company, Her Majesty's Inspectors of Salmon Fisheries Spencer Walpole and Thomas Henry Huxley, and Sir Francis Burdett, plus at least a dozen MPs, innumerable knights, ladies and members of the officer class. The ghost at the feast, and the binding force that had brought them all together, was Francis Trevelyan Buckland. The sole business of this impressive gathering was to discuss proposals for – at last! – a Great International Fisheries Exhibition to be held in London in the following year, 1883. It would be the biggest and best exhibition of its kind ever held anywhere in the world.

The Prince of Wales was loudly cheered during an ebullient speech which, in substance if not in style, might have been scripted by Frank himself. His Royal Highness had many people to thank, but he saved the most important until last.

Before I sit down there is one name which I cannot help alluding to, and it is with feelings of sincere regret that he is no more among us to give us his valuable assistance. The loss of so distinguished a man as Mr Frank Buckland, whose thorough and scientific knowledge of everything connected with fish and fishing must have been of the highest value, cannot be but deeply felt by all.

'Hear, hear,' said the great and good in their multiplicity of accents. It might have been thought a further compliment to Frank that he should have been succeeded as inspector by such a distinguished scientist as Thomas Henry Huxley, though in fact Huxley turned out to be something of a disappointment. He was a fine anatomist, evolutionary theorist and champion of scientific education, but he signally lacked Frank's hands-on, salmon's-eye

approach to the study of nature. 'Since Mr Buckland died,'
J. W. Willis Bund would complain in *Salmon Problems* in 1885,
'it seems to be no one's business to collect the various facts relat-
ing to the Salmon that are reported from time to time.' The
Times obituarist was right. Frank's death would 'leave a blank
not easily filled up'.

Many more meetings ensued, and in June 1882 the organising
committee set itself up in offices at 24 Haymarket. But while
London went on talking, Edinburgh preferred to act. In April
that year, independently of London's royal powerhouse, Scotland
held its own National Fisheries Exhibition in Waverley Market.
Prominent among the organising committee was Archibald
Young, who had visited the exhibition at Norwich. It was a magnificent effort. Though declaring it 'subordinate to the Exhibition
to be held in London next year', *The Times* was impressed.

The vast hall . . . has been transformed into a gay and lively
fancy fair, crowded with objects of interest and adorned
with bright and appropriate decorations. Seven parallel
rows of tables run from one end of the hall to the other,
affording 2,100ft of lineal space for the display of the multi-
farious exhibits, not to mention the broad areas under the
galleries which are occupied with models of fish ladders,
fish passes, salmon bags, nets, and the apparatus for fish
culture. The galleries, the girders, and the pillars that support the roof are festooned with herring nets, buoys, vases
of goldfish, lifeboats, and many forms of tackle.

The whole thing might have been dictated by Frank. It was reported
afterwards that the exhibition had attracted 136,000 paying visitors and taken 'the very respectable sum of £5,236'. The gay and
lively fancy fair, however, was about to be crushed into a historical

footnote. The Great International Fisheries Exhibition of 1883 would bear the full imprint of Britain's imperial greatness, a triumphant reiteration of global power, an event to amaze the world. A correspondent in *Science*, the journal of the American Association for the Advancement of Science, described the full panoply of Britannia at the top of her ceremonial form. On opening day:

> The entire English court was present; and the gorgeous costumes of the royal family and their households, the picturesque garments of the foreign ambassadors and commissioners, the military and naval officers, the yeomen of the guard, the Queen's watermen, the English, Scotch and Irish fishermen, the fishwives from Scotland, Belgium, France, and Holland, mingled with the bright decorations and the striking objects among which they were passing, made the scene very brilliant and impressive.

He forgot to mention the state trumpeters, the forty-one-gun Royal Salute in Hyde Park, the orchestra and 400-voice choir, plus enough feathered helmets to denude an entire continent of ostriches. The band of the Grenadier Guards performed every day (if fine in the gardens, if wet in the Inland Fisheries Promenade). The only disappointment was that 'under the injunctions of the Royal physicians', the Queen was unable to perform the opening ceremony, so the Prince of Wales had to deputise. The exhibition site was immense, taking up twenty-two acres of the Royal Horticultural Society's grounds at South Kensington,* where it had taken just six months to put up all

---

* The society moved from the site in 1888. It is now shared between the Science Museum, Imperial College and the Royal College of Music.

Thirty-one countries, including the United States, were represented at
The Great Exhibition

the buildings.* It was a monster version of Frank's Museum
of Economic Fish Culture, covering all the same issues – fish
culture, and every aspect of sea and freshwater fishing – but on
a grand scale and with displays from thirty-one countries and
colonies. The United States alone contributed £10,000 to the
budget and provided a fishery steamer for the exhibitors. There
were aquaria, fishing tackle, diving and lifesaving equipment,
lifeboats and demonstrations of fish cookery. Not least of the
attractions was the extensive use of electric lighting, which
excited much comment in *The Times*.

---

* Not everyone welcomed the changes. On 30 January 1883, a correspondent complained in *The Times*: 'Thus, the gardens which were the pleasantest
resort of West-end people are now being covered with buildings . . . and the
flower-beds and parterres at which horticulturalists pursued the study of their
favourite science are now covered with timber and iron.'

On Monday 18 June the Prince of Wales opened the first of several conferences organised to coincide with the exhibition. The inaugural address was by Thomas Henry Huxley, and it is worth reflecting on what he said. Specifically, he gave his full and unequivocal assent to Frank's opinion that stocks of herring and other species were inexhaustible and could not be overfished. *The Times* took particular note of his observations about cod off the coasts of Norway.

In January and February . . . the fish form a 'cod's mountain' of a depth varying from 120 to 180 feet. A square mile of such a shoal . . . would contain no less than a hundred and twenty millions; and this, be it remembered, is but a portion of one of the shoals that haunt one single corner of the sea. Again, the cod must be fed; and what they feed on is chiefly herrings. A cod would be hardly likely to confine himself to one herring a day, yet if he did, the week's total for the one shoal of which we are speaking would amount to 840,000,000, or rather more than double the whole number of herrings which all the boats of all the Norway fishermen bring to land in a whole year. In other words, the supply of such fish as this is practically inexhaustible; and however hard we fish we are not likely to make any difference whatever in the supply that the sea is for ever producing.

Thus it follows that if G. H. O. Burgess is to convict Frank of 'pseudo-science', then Thomas Henry Huxley must be convicted of the same offence. Frank knew better than anyone that his knowledge was incomplete. His most insistent call had been for more and better science, and in particular for a better understanding of the life cycles, behaviour and vulnerabilities of fish. Now others were taking up the cause. In September the General Committee of the British Association reflected that one of the most remarkable results of the fisheries exhibition was the unavoidable conclusion that 'next to nothing' was known about the lives of fish. This appalling state of ignorance, for so long lamented by Frank, could not be allowed to continue. Frank himself had appealed for government-funded research into 'the dark and mysterious habits of food fishes' as long ago as 1867. Finally stirred from their slumbers, the learned gentlemen declared a suddenly urgent need for a marine laboratory to do exactly what Frank had suggested. Only then would it be possible 'to say when fishing should and when it should not be allowed; what nets were injurious to the fish, and what should be tolerated; what amount of fish should be taken, and so on'. Two weeks later, while the exhibition was still running, a circular proposing the establishment of 'a marine observatory on the British coast for the study of marine animals and plants' was signed by, among others, William Henry Flower, Hunterian Professor of the Royal College of Surgeons, who would succeed Richard Owen as director of the Natural History Museum; Richard Owen himself; Philip Sclater, secretary of the Zoological Society of London; and distinguished professors of zoology, natural history, anatomy and physiology at the universities of Cambridge, Oxford, London and Manchester. The one notable absentee was Thomas Henry Huxley, whose unshakeable belief in the infinite abundance of the sea meant that, in his opinion, there was nothing for such a laboratory to investigate.

When the exhibition closed on 31 October it had received a total of 2.6 million visitors, at an average of 18,545 per day. Few would quarrel with *The Times*'s verdict on it as 'the most successful show in the world', and Frank must have been pirouetting in his grave. Yet there was a darker side. Through its very size and magnificence, the exhibition had shifted attention and resources away from its inspirational and far-sighted parent, Frank's Museum of Economic Fish Culture. On 25 April 1883, as work on the exhibition neared completion, the chairman of the Tyne Salmon Conservancy, John M. Ridley, complained to *The Times*:

> Pray allow me . . . to call the attention of those responsible for the preservation of the Buckland Fish Museum at Kensington to the disgraceful state of untidiness, dirt, and neglect it is in. It will appear a gross incongruity to the visitors of the Fisheries Exhibition that the only collection of its kind in this Kingdom should be allowed to fall into decay . . . It will also seem a poor return by the nation to the donor, who has done more practical work towards the preservation of our fisheries than all the scientists in the world, thus to show our ingratitude for his services. In America and on the Continent his name is held in great honour and his services in the cause of pisciculture are more appreciated than in his own country, though along our coast no name is more fondly remembered by our hardy fishermen than that of the genial Frank Buckland.

A bandwagon had now started to roll. In the driving seat was Ray Lankester, Professor of Zoology at University College London and another future director of the Natural History Museum. He was a large man with a big mouth, of whom it would be said that he did not suffer fools gladly – another way of saying he was better at making enemies than friends. He could also be wrong. When the first accurate depiction of the previously unknown okapi was sent to London by the explorer-naturalist Sir Harry

Johnston, Lankester denounced it as a hoax. In the cause of fishery science, however, he was firmly on the side of the angels and not afraid to stand up to Huxley. It would take time, but all the pieces in Frank's vision gradually would fall into place (though not everyone would remember or acknowledge whose vision it had been).

On 30 March Lankester and his supporters met at the Royal Society and formed themselves into the Marine Biological Association. Lankester by now had persuaded Huxley to be the association's first president, while Lankester himself was secretary. By October 1884 they had secured a prominent site on Plymouth Hoe, where Britain's first on-shore marine laboratory would be opened on 30 June 1888. It remains a world-leading centre for marine biological research, and has advanced the careers of at least a dozen Nobel laureates. This is pretty much the point at which science says its final thank-you to Frank Buckland and waves him a fond farewell. The infinity-of-the-oceans school of thought had not yet entirely died out, though it was drowning in a sea of contrary evidence. Huxley was an early casualty. In 1889 he was so appalled by a plan to study the effects of trawling in the North Sea that he resigned the presidency in a huff. The study went ahead from a new base at Grimsby. And so, over the decades, it has continued. From the tiniest and most unregarded of minnows, marine science has grown into one of the world's most important research disciplines, delving microscopically into every physical, chemical and biological quirk of the oceanic environment. From our own Bucklandesque perspective there is much to celebrate. Through the continuing work of the Marine Biological Association and the UK government's Centre for Environment, Fisheries and Aquaculture Science (Cefas) at Lowestoft, Britain is now a world-leading authority on every kind of marine resource from plankton to wind power, oil and gas. On this score at least, Frank's shade can rest in peace. No policymakers anywhere in the world can now plead ignorance as an excuse for their mistakes.

The review of George Bompas's biography of Frank, which filled page 4 of *The Times* on 27 May 1885, began with a mock complaint:

> The fault we find with Frank Buckland's *Life* is that the early chapters are too delightful. The volume is full of instruction and varied entertainment for all who sympathize in Buckland's favourite pursuits; but the story of the boy as father of the man is so piquant and original as rather to spoil us for what is to follow. We can recall no equally striking example of the precocious bent of irrepressible instincts.

Frank lived on in other ways too. In the years after his death, the chief guardian of his reputation was Henry Ffennell, son of the former fisheries inspector William. On Thursday, 26 November 1885, Henry was the bearer of sad news. The last of Frank's pet monkeys had died. The passing of 'Tiny the Second', he told readers of *The Times*, meant that only one member of the Albany Street menagerie now survived, a female parrot struck dumb by grief. Tales of Frank's pets, however, were always apt to confuse. When the parrot finally fell from its perch in August 1890, 'Tiny the Second' was revealed to be still alive (she survived until January 1895).

Frank's books lived on too, and new editions continued to appear. (In November 1886, Smith Elder & Co.'s new popular edition of *Notes and Jottings* was advertised at five shillings.)

But the pitiable condition of Frank's most tangible gift to the nation, the Museum of Economic Fish Culture, continued to horrify everyone who knew its value. In October 1887 a 'learned professor' wrote in anguish to Henry Ffennell: 'If England is so poor that her scientific collections are to be second rate, let them be second rate; but if she is to have scientific collections at all they should not be housed in, or approached through, filthy, repulsive, unwholesome sheds.' In a letter to *The Times*, Ffennell complained that rain was dripping through the roof so that specimens had to be draped in tarpaulin.

> The whole place presented as demoralizing an aspect as it is possible to conceive, and I venture to think that such a state of things would not be allowed to exist in any similar institution in the world; not even in the very smallest sub-sidized museum either at home or abroad . . . After sudden and heavy falls of rain a copious flow of water comes in at several points, and it is not at all easy to pick one's way without getting wet feet.

Frank's life's labour had been given to the nation under trust, and that trust had been disgracefully betrayed. It was a theme to which Ffennell would return time and again, never more furiously than in October 1892 after he had taken a short cut over some wasteland near Queen's Gate. There on the ground lay a substantial part of Frank's collection . . . 'Models, casts, preparations in bottles, &c., all lying higgledy-piggledy . . . as if they had been cleared out as so much rubbish to be got rid of as soon as possible'. The decline was irreversible. In 1898 a House of Commons select committee declared the collection to be obsolete and recommended its abolition. A committee of the Piscatorial Society, while agreeing that the museum was suffering gravely from neglect, argued that it could still be brought up to date and honour Frank's intentions. Along the way it dug up something even murkier than the cavalier disregard of a unique resource. Where was the money Frank had left in trust to the director and assistant director of the South Kensington Museum to pay for lectures on fish culture? The committee looked, but 'failed to ascertain what has been done with this money. All that they know is that no such lectureship exists.' This caused outrage among fishery boards, angling societies and leaders of the fishing industry. In March 1899 a petition against the museum's closure seemed to have been signed by every duke, marquess, earl, knight, MP, QC, judge and professor who had ever held a fishing rod or enjoyed a dish of salmon. They were powerful voices, and in principle they were right. But the pragmatists were right in their way too. The Museum of Economic Fish Culture had rotted beyond all hope of salvation.

References to Frank and his work continued to appear throughout the 1890s. In November 1895 a correspondent in *The Times* wrote in laudatory terms of his campaign against salmon poachers. In January 1901 another hailed him yet again as the man the whole world had to thank for artificial fish hatching. In 1905 an advertisement for St George's Medical School named him alongside John Hunter and Edward Jenner in a list of 'great men of St George's and what they have done for the nation'. Not even the death of Henry Ffennell in 1909 could stem the flow. In 1910 John Upton's *Three Great Naturalists*, with its potted biographies of Charles Darwin, Frank Buckland and the Reverend J. G. Wood, was published by the Pilgrim Press. The *Spectator* applauded Upton's choice of subjects, but suggested it was a huge mistake to present them as equals. 'Darwin made an epoch in natural science; the other two were not men of genius, but they were workers of the first class.' This was only half right, and may have been the first sign of history beginning to get Frank wrong. While it was unarguably right for the *palme d'or* to be handed to Darwin, Frank deserved better than to be handcuffed to John George Wood. The reverend gentleman was a talented populariser whose *Common Objects of the Country* was a runaway bestseller, but he was no more than that. Frank more than equalled him in popularity but greatly exceeded him in scientific accomplishment. It is with Darwin, not Wood, that Frank properly should be compared, and the comparison may not be as odious as it seems. While Darwin was the greater intellect and has had a far more profound influence on our view of the world, it could be argued that Frank's work on fisheries was of more direct public benefit. And yet, despite Upton's best efforts, he was gradually fading from view. His *Encyclopaedia Britannica* entry in 1911 was both short (barely two column inches[*]) and turgid: there was space only for a brief résumé of his education, his medical and scientific appointments and an incomplete list of his publications. His accomplishments were summarised as 'a good deal of out-of-the-way research in zoology' and 'being largely responsible for the increased attention paid to the scientific side of pisciculture'. There was no mention of his museum. As Frank by now had been dead for thirty years, there were few people who still remembered him. Even so, 1912 brought a further rash of articles and letters to *The Times*, invoking him as the pre-eminent authority on fish, and most especially on heavyweight salmon. In May the auctioneers Knight, Frank & Rutley brandished his name

---

[*] His father, William, got eleven inches.

in an advertisement for a Surrey mansion whose lakes he had stocked with trout. Thirty-two years after his death, his name still had selling power. As late as 1914, his story of the tortoise-insect at Southampton railway station could still be alluded to as 'famous'.

Frank vanished from sight during the First World War, but was resurrected in April 1918 as an expert witness in a controversy over the edibility of salmon kelts. A year later he was enjoined in a campaign to persuade the government to create a Ministry of Fisheries (the Ministry of Agriculture and Fisheries was established later in the same year). In September 1919 *The Times* was still citing him as the pre-eminent authority on oysters. In August 1921 its obituary of Lady Shelford – 'A Lady of the Old School' – made much of her long-ago brush with celebrity.

She remembered well Frank Buckland, the naturalist, and told of his coming into church one Sunday with an enormous giant on one side and the smallest dwarf of the day on the other. She was persuaded by Buckland to construct a piscatorium in her bedroom when she was busy with her trousseau.

This cherished memory was trotted out *forty-one years* after Frank's death, though references to him by now were infrequent. Old arguments still rumbled on. One of the most eminent marine scientists of the day, Sir William McIntosh,* continued to maintain that Frank and T. H. Huxley had been

---

* Commissioner of Fisheries for the Dominion of Canada; Professor of Zoology, St Mungo's College, Glasgow; president of the Andersonian Naturalists' Society, Glasgow.

right – trawlermen could no more empty the sea than they could exhaust the air by breathing it. But with increasing speed the pendulum was swinging the other way. In September 1922 a three-day conference of the British Association at Hull declared itself convinced by the evidence. In the case of flatfish at least, the fishermen were progressively depleting the stocks on which their livelihoods depended, and were having to sail further and further to net smaller and smaller catches. Then, as now, commercial and scientific interests viewed each other with suspicion. As reported by *The Times*, the conference revealed 'how much research has yet to be carried out, before agreement of experts convinces commercial men that their advice is worth following'. The best tool for such work, it was agreed, was fishery research vessels – an idea put forward by Frank in 1867.

Still he was not quite forgotten. A ten-foot model of a sailing trawler, exhibited at the Science Museum in 1926, was named after him and apparently much admired. This was a memorial of sorts, but four years later came a commemoration of much greater and more lasting significance. At last, fifty years after his death, the money Frank had left in trust for lectureships was to be put to its intended purpose. The bequest was not quite intact: early on, one of the fund's administrators had embezzled £1,000 of the £5,000 Frank had left, but the residue was enough to set the reinvigorated Buckland Foundation on its feet.* In 1930 it awarded its first annual Buckland professorship to Walter Garstang, Professor of Zoology at the University of Leeds, whose inaugural lecture, 'Frank Buckland's Life and Work', reminded his listeners – few if any of whom could have had any personal recollection of their benefactor – of exactly

---

* The foundation carefully invested the money to keep pace with inflation. Even today, grown to six figures, it still retains its value.

why they were there. He was succeeded in 1931 by William Leadbetter Calderwood, director of the Marine Biological Association laboratory at Plymouth, whose subject could hardly have been closer to the founder's heart: 'Salmon Hatching and Salmon Migrations'. Professors have been appointed, and lectures delivered, almost every year since (there was a break for the war, and a few gaps in the 1970s) on subjects touching upon every aspect of commercial fishing, from the generalities of supply to the specifics of particular species and their environments. In 1964 the professor-lecturer was none other than Frank's biographer, G. H. O. Burgess, who spoke on 'Developments in the Handling and Processing of Fish'. As time passed, the issues became ever more complex and the hazards to fish stocks ever more acute. Contributors more recently have had to reflect on the further complication of the science-versus-commerce imbroglio by the intrusion of international politics. One would like to have heard Frank's opinion of the European Common Fisheries Policy and all the consequent lunacies of discards, 'black fish' and competitive quota-mongering.*

Recollections of Frank continued to surface occasionally in the 1930s. In 1933 a dispute about the British sturgeon record was settled by a *Times* correspondent who cited a fish caught on the Wye in 1877. 'As Mr Frank Buckland conducted a *post-mortem* on this specimen these figures may be taken to be authentic.' For a few very old men with long memories, it might have awakened an ancient echo. *Because*

---

* The effect of political regulation was the dominant theme of 2015, when the joint professors were Colin Bannister, former senior fisheries science adviser at Cefas ('Has EU Fisheries Management Achieved the Recovery of Depleted Stocks?'), and Carl O'Brien, chief fisheries science adviser to Defra ('Future Fisheries Management and Governance in the EU').

Buckland says so. In 1935 a writer retold the story of Frank's misadventures with porpoises, embellished with lines from Thackeray's elegy. When the Piscatorial Society celebrated its centenary in 1936, it congratulated itself for having been able to count Frank among its members. In September 1937 the prolific English writer E. V. Lucas managed to fill a whole column in the *Sunday Times* with an account of Frank's hedgehog-versus-snake experiment, in which '[Frank] tells us not only that the hedgehog eats snakes but that it can run faster than a snake can proceed. If Buckland says so, this must be true.' In a letter to the same paper, an M. E. Durham of London NW3 described how his father* had taken him to Albany Street to watch Frank cast a 70lb salmon. Small wonder the memory had never left him.

> Buckland greeted us. He was coatless and bare-armed. In the middle of a long table was a tray with a dissected animal on it. At the end of the table was a large bowl of Irish stew. Buckland was running up and down, now taking a cut at the dissection and now gobbling a spoonful of the stew. 'Have some?' he asked, hospitably waving towards the stew . . . Buckland seized a stick and poked behind a bookcase and out rushed a hare . . . He showed us the original curari poison brought by Bates from the Amazon, put it in my hand, and said I was holding enough to poison half London.

Reflecting on his 'red-letter day', Durham expressed the rather forlorn hope that 'there are children to-day who read Buckland'.

After this, references to Frank became sporadic and mostly trivial. In July 1939, *Because Buckland says so* was the clincher in

---

* Probably Arthur Edward Durham, consultant surgeon at Guy's Hospital.

a long-running argument about the mythical 'frog showers', in which juvenile frogs, supposedly sucked up from ponds, were believed to fall with rain. (The truth, which Frank was one of the first to articulate, is that frogs and toads travelling overland will conceal themselves during dry weather and emerge in hordes after rain, creating the impression of a sudden 'shower'.) In June 1942, in response to the disbelieving mirth of MPs who doubted that anything worthwhile could ever have been caught in the Serpentine, *The Times* reminded them that Frank in 1869 had drained the water and removed eleven cartloads of fish. More pertinently, in July 1948, the distinguished marine biologist C. M. Yonge celebrated Frank's pioneering work with freshwater hatcheries. Slowly, slowly, the world was catching up. In September 1951, the capture of some tagged fish near Yarmouth confirmed his prediction that sea trout caught off East Anglia were spawned in the rivers of Northumberland. In 1954 his name was drawn into a protest about the poor condition of Brighton Aquarium, and in 1955 Dr Denys Tucker[*] wrote to the *Sunday Times* from the Department of Zoology at the British Museum, quoting him in an ongoing correspondence about the diet of nineteenth-century apprentices.[†] Apart from the references to raw-onion soporifics on the Internet, this is the very last example I can find of *Because Buckland says so.* He reappeared briefly in 1963 when Professor Yonge reprised his

---

[*] A controversial character, sacked from the Natural History Museum in 1959 for claiming to have seen the Loch Ness Monster.
[†] The mythology was that the apprentices were exceptionally well treated because of a stipulation in their indentures that, in Frank's words, they 'should not be bound to eat salmon more than a certain number of days in the week'. This was indeed a mercy, for the salmon in question were not plump, healthy fish in prime condition but scrawny, spawned-out kelts which, as Frank himself could attest, were peculiarly nasty to eat.

366 THE MAN WHO ATE THE ZOO

account of a white whale arriving at Westminster Aquarium in 1878; and, more bizarrely, when he cropped up in the *Times* Christmas Quiz. Question: 'Who mesmerized Tiglath Pileser in the Botanic Garden in 1847?'

Four years later came G. H. O. Burgess's authoritative biography, *The Curious World of Frank Buckland*, which I believe did not get the readership it deserved. And that, pretty much, was that. In its 'On This Day' feature on 14 December 1990, *The Times* reprinted a letter of Frank's, written from the Athenaeum in 1866 on fish ladders in the Thames; and on 13 August 1994, it reproduced a short piece which Frank had written for *Land and Water* in 1877 on the Birds Preservation Act. One final joke – question 4 of *QI*'s £1,000 prize E Quiz,* which appeared in *The Times* on Saturday, 29 December 2007:

What did the zoologist Frank Buckland describe as 'horribly bitter'?

Answer: earwigs.

The tough little fishing port of Anstruther stands, four-square to the bone-chilling easterly winds, on the North Sea coast of East Fife, twenty-five miles across the Firth of Forth from Edinburgh. Six and a half miles to the north-east is the county's easternmost extremity, Fife Ness, a jutting elbow with St Andrews just round the corner. Once it was fish that kept the local economy afloat; now it's golf, sailing and tourism. On a cold January morning, the view from the harbour wall is still a close match for the nineteenth-century photographs that adorn so many local walls. The quayside buildings are recognisably the same, though their functions now owe more to the tourist than to the herring (fish-and-chip shop, restaurant, wine bar, betting shop, Turkish

---

* Set by the *QI* BBC panel show chaired by Stephen Fry.

takeaway, coffee shop, ice-cream parlour). Masts still cluster in the stone-walled inner harbour. But there is a critical difference. Apart from a single rank of inshore fishing boats lolling beneath the harbour's southern wall, the masts all belong to leisure boats – yachts and cruisers – maritime diversions for people whose wealth owes little or nothing to the sea. The old sailing luggers, herring drifters and whalers survive only in sepia. Even the few working boats, landing lobster, crab and prawn, are serving very different markets from the old herring fleet. Frank in his prime, when the place handled more fish than any other port in Scotland, would have loved it. Today it would both sadden and gladden him. The sad bit is what I can see in the harbour: the ghost of Anstruther's past. The gladness has to be sought.

At the harbour's north-west corner is what looks at first like a tea shop – unsurprising, as that's exactly what it is, a warm and welcoming cafe where shoppers meet for coffee, and which is packed at lunchtime by high-school girls in search of soup and paninis. But all is not as it seems. To reach the counter you have first to pass through a souvenir shop, which opens onto a cobbled courtyard. Around and beyond the courtyard is a muddle of historic buildings – a fifteenth-century chapel, a sixteenth-century abbot's house, an eighteenth-century merchant's house, an old chandlery, some cottages, a boatyard, the once-upon-a-time offices of the Anstruther Whaling Company, a former tavern and a smokehouse – all knocked through in a disorientating maze of interconnecting spaces. Corridors lead into rooms that spill into ever larger galleries and then, stunningly, into vast halls the size of churches. You have entered one of the great glories of Britain's eastern seaboard, the Scottish Fisheries Museum.

It was opened with modest space and resources in 1969, but has grown to become one of the biggest museums dedicated to a single industry anywhere in Britain. Here you will find everything from fish hooks to complete fishing vessels. There are life-size waxwork tableaux of a fisherman's cottage, a fish-gutting shed, a fishmonger's shop; historic photographs and paintings; the actual bridge and galley of a trawler; exquisite models of fishing boats through the ages; the working boatyard where shipwrights bring historic vessels back to life; every kind of fishing tackle ever used to catch every known species of marketable fish; a whaling gallery and, most poignantly, a tiny chapel whose walls are covered in brass plaques, more than I can count, one for every Scottish fisherman who has been lost at sea since 1946. Life for a fisherman may be safer than it was in the nineteenth century, but it doesn't hurt to be reminded that fishing is still one of the hardest and most dangerous ways to make a living, and that human lives are the real cost of fish.

I find what I am looking for in the 'Fishing into the Future' gallery. Tucked away in a corner is a single display case containing a marble bust, some plaster casts and a three-pronged fish spear. This is it. Acquired by the museum even before it opened, this small illuminated shrine is all that remains of Frank Buckland's Museum of Economic Fish Culture. The bust is the one commissioned by the Buckland Memorial Fund in 1881. The casts – an octopus, a salmon, a swarm of baby sharks – are, as far as I know, the only ones Frank made that are on public display anywhere in the world. Upstairs in the storerooms my guides, John Firn and Richard Shelton, respectively clerk and chairman of the Buckland Foundation, which has its headquarters here, manage to dig out a few more: a strange, pot-bellied 'Deformed trout', signed by Frank and dated 21 April 1871; a mysterious and pleasingly ugly 'Arctic chimaera' from 1868; a well-named 'Monster garpike'; an American black bass; a sole

Frank's marble bust and plaster casts at the Scottish Fisheries
Museum – the last remnants of his collection on public display

and a dab mounted in a strangely intimate embrace; a brown
trout with 2,470 salmon eggs in its stomach; and numerous
salmon. Altogether the museum holds forty-five of Frank's casts.
There is also a stuffed otter, a turtle shell, narwhal tusks and
a large amount of fishing tackle: hooks, spears (including some
confiscated from poachers), a scallop dredge, a scalloping harrow, a fish drag, a mesh gauge. Some of these are displayed in
other parts of the museum. Most are in the stores.

So there they are: like a fading echo, the last few scraps of a
great man's life. Burgess believed that fate had been unkind
to Frank, and he was right. He died too soon. Had he lived
long enough to see his ideas come to fruition, he would surely
have been given the credit, and the place in history, that he
deserved. Instead, his name has been all but expunged from the

record. There is no room for him in *Encyclopaedia Britannica* now, and few historians of the nineteenth century deem him worthy of mention. This is not just a pity. It is distortion by omission, a travesty, an injustice. Burgess was probably right to conclude that Frank was not a truly great scientist, though I quarrel with his assertion that he was 'not really a scientist at all'. He was through and through a man of science, one who enquired, observed and recorded with a scientist's obsession for detail and respect for the facts. His conclusions, right or wrong, were invariably grounded in the best evidence available to him. That the evidence was so often inadequate was the very issue that propelled him to champion fishery science, which is why later, better-informed generations are so heavily in his debt. As Burgess aptly put it, he prepared the soil which others would sow and reap. History has not repaid him with the generosity that he bestowed on others. Ironically, the very warmth of Burgess's tribute hints at the prejudice that would condemn Frank to his posthumous gulag of obscurity.

His life was to him an amusing adventure, shared with thousands of others who were the richer for his writing; they laughed with him, they enjoyed his jokes, they were interested in what interested him. And when he died they lost a friend and a companion and were so much the poorer.

Frank Buckland committed three unforgivable sins. He was popular; he repudiated Darwin; and he did not live long enough to change his mind, as he surely would have done, about the infinite capacity of the sea. The worst sin was the first. Serious scientists were not supposed to be *popular*; they did not entertain people or make them laugh. Serious science spoke a special language and wore a furrowed brow. Well, let the pedants follow their well-trodden path to the *Oxford English Dictionary*, and there reflect on the difference between seriousness and solemnity. Frank put a smile on science's face, and it would be good to see it put back there.

Afterword

If I felt any regret during the writing of *The Man Who Ate the Zoo*, it was that I was unable to flesh out the character of Hannah Buckland. She emerges from her husband's story as a supportive, perhaps even long-suffering wife who, by the standards of the time, was well treated by the father of her illegitimate son. Frank's marriage to her *after* the premature death of 'poor little Physie' is a testament at least to Frank's sense of responsibility – by no means the norm among Victorian men who impregnated socially inferior girls – though there is little suggestion of romantic love. 'My Dear Wife' and 'Your affectionate husband' are fairly low down on the spectrum of passion. In return for his affection, Hannah seems to have been a more than usually supportive wife unfazed by Frank's erosion of the boundary between 'home' and 'menagerie', or 'order' and 'chaos'.

I had not known when I wrote the book that Hannah had a sister, Emma, whose life ran on a similar track to Hannah's own. According to her great-granddaughter Mary Pritchard, who has kindly written to me about her, she too 'married well'. Given the family connections, it was perhaps no great surprise that she should marry another contributor to *Land & Water*, Charles Bayly, and that she should live near her sister in Albany Street. It is to Emma that we owe the survival of what, to modern eyes, is an extraordinary letter from Frank to his wife, written on the headed notepaper of the Salmon Fisheries Office, 4 Old Palace Yard,

Westminster, SW. It is addressed to Mrs F Buckland, 37 Albany St, Regent's Park, dated 2 March 1872, and marked 'Private'.

My Dear Wife,

I have this morning dictated the terms of my will to my friend & legal adviser Bennett.

I think it right to tell you that I have left you everything – except my museum – all my property in fact – between six & seven thousand pounds altogether.

In the case of my death you will have about £400 a year. If you wish to leave me you will have the same. I have so ordered the will that if you marry after my death your future husband will not be able to rob you. I wish you to put this letter carefully away in case my will is disputed. I hope to sign it next Wednesday.

Your affectionate husband
Frank Buckland

Show this to no body

It is impossible to know what was the likelihood of Hannah feeling any 'wish to leave' her husband, but it is hard to believe that this was the kind of dispensation commonly extended to Victorian wives. Questions abound. Frank did nothing without a reason. Perhaps he did not want her to feel, and did not want to feel himself, that she was bound to him by money alone – a somewhat backhanded declaration of love. Perhaps he *wanted* her to leave. One might wonder why no more children were born after the loss of Physie. Or perhaps there was something else on his mind altogether. Whatever it was, there is no evidence that Hannah ever took advantage of the offer. Neither is it clear why

Frank feared the will might be challenged. By whom? There were no dependent relatives, and no evidence of creditors. Perhaps the subsequent embezzlement of part of the sum he had left in trust to the Director and Assistant Director of the South Kensington Museum indicates a possible target for his suspicions. We will never know.

I thank Mary Pritchard also for her revelation that the family's 'animal eccentricity' did not die with Frank in Albany Street. In a distinct echo of Christ Church pond, Mary's own father and his twin sister kept the tradition going with pet alligators. I am grateful also to Frank's great-nephew, the author Roderick Gordon (great-grandson of William Buckland's daughter/ biographer Elizabeth Oke Gordon and, as it turns out, a near neighbour of mine in Norfolk) for revealing the existence of his own collection of papers and memorabilia from his great-great-grandfather William Buckland, Frank's father. Here again, family traditions clung on. From a very early age, Roderick was taken by his father on Buckland-style fossil forages along the south coast, and regaled with stories of Frank's singular tastes in meat and fish. Influences from William and Frank's lives duly found their way into Roderick's *Tunnel* series of subterranean adventures. What better name could there have been for a strange underground race than William Buckland's gift to the scientific lexicon, 'Coprolites'?

List of Illustrations

Acknowledgements

There is no better way to reassure yourself of the innate generosity of your fellow humans than to embark on a book and ask for advice. Even after long and grateful experience I continue to be amazed by the lengths to which people will go to help a complete stranger find a reference, confirm a fact or locate a picture. Some of those who gave their time were lifelong Buckland enthusiasts; others had never heard of him. It made no difference to their willingness to help, and this book would have been a lot more difficult to write, and a lot worse in its execution, without them. I would like to thank in particular:

Michael Hellyer, Archivist, St Martin-in-the-Fields; Suzanne Foster, Archivist, Winchester College; Emma Anthony, Project Archivist, and Fahema Begum, Assistant Archivist, Royal College of Surgeons; Linda Fitzpatrick, Curator, Scottish Fisheries Museum; John Firn, Clerk, and Richard Shelton, former Chairman of Trustees, the Buckland Foundation; Sarah Pearson, Carina Phillips and Bruce Simpson, Curators, Hunterian Museum, Royal College of Surgeons; The Very Revd Dr John Hall, Dean of Westminster; and Oliver Riviere, without whose expertise I would have been unable to reproduce many of the pictures.

My thanks go to my agent, Jonathan Pegg, for a magnificent lunch, limitless enthusiasm and bold encouragements; to Poppy Hampson at Chatto for commissioning and editing the book with all her customary skill and sensitivity; and to my copy-editor, Katherine Fry, for her work with dustpan and brush, whisking away repetitions, obfuscations and howlers.

Of my wife, Caroline, I can say only that she is a gift far beyond the scope of thanks.

Index

Page references in *italics* indicate photographs or illustrations.
FB indicates Frank Buckland.

EATS VARIETY OF ANIMALS
11–12, 14–15, 174–5; bear
14–15; boa constrictor 117; bull
bison 279; crocodile 11–12, 14,
35; deer 174; earwigs 1, 366;
eland (antelope) 145–6; frog,
swallows live 116; frogs, love of
eating French 116; giraffe 161;
hedgehog 44, 150–1; horseflesh
269–71, 279; kangaroo 174–5;
kelt 187, 252, 252n; lumpfish 72;
monkfish 191; oyster, extremely
large 187–8; panther 56, 159;
rhinoceros 295; swordfish 282–5;
tripan 174; tuna, possible tasting
of 281–2; turtle 72
EVOLUTION, REACTION TO
IDEA OF: adaptation as proving
perfection of God's Creation,
views 284–5; Creationist faith
and 9, 71, 119–20, 153, 165–6,
256, 284–5, 297–8, 306, 313, 317,
329, 345; 'Darwin's tubercle' and
304–5; Darwin/Darwinism,
reaction to 8–9, 75, 165–7, 185,
275, 284–5, 297, 298, 304–5,
312–13, 344, 345, 359–60; deer
breeding and 312–13; mole
eyesight and 317; serpents in
Garden of Eden, thinking on
119–20
FAMILY: brothers 31, 34, 46, 58;
father and see Buckland, Dr.
William; grandparents/ancestry
6, 29–34; marital home 179,
184–9 see also Albany Street,
London, FB house at; marries
76–9, 177–8, 261; mother and see
Buckland, Mary; sisters 5, 6, 7,
9, 18–19, 20, 21, 46, 53, 57, 58,
60, 62, 78, 80–1, 82, 101, 102,
347; son see Buckland, Francis
John; wife see Buckland, Hannah
FINANCES 77–8, 79, 80, 89, 177,
213, 362
FISH/SEALIFE, INTEREST IN:
Australia and New Zealand, role
in introduction of salmon and
trout into 81, 146, 199–200, 285,

286, 326, 327; books and see
under individual title of work;
Brighton Aquarium, role in
establishing 290–1, 292–3,
294–5, 296, 365; British
Fisheries Preservation Society,
joins council of 180; Creationist
faith and 153–4; De Garris v.
The Mercantile Marine Insurance
Company (1868), testifies in case
of 282–3, 285; dogfish, interest
in 153–4; English inertia with
forward-thinking attitudes of
other nations, contrasts 290;
fishmongers, tolerance of sharp
practice amongst 154–5; focus
on dependable supply of affordable fish for ordinary people
179–80; George Butler
('Robinson Crusoe') and 189–91,
191, 193–4, 324–5; growth of
interest in 151–5; hatching of
fish, interest in artificial 19,
179–84, 183, 194–8, 199–200,
210–11, 213, 216, 252, 259–60,
261, 263, 278, 286, 299, 359,
363, 365; Inspector of Salmon
Fisheries role see INSPECTOR
OF SALMON FISHERIES;
instinct for water, interest in
127–8; management of fish
stocks/overfishing, attitudes
towards 289–90, 314–15, 330–41,
353–4, 356, 361–2, 363, 363n;
Museum of Economic Fish
Culture, Kensington see Museum
of Economic Fish Culture,
Kensington; ova collection and
transportation 181–4, 197,
199–200, 216, 247, 259–60, 285,
326–7; oysters/oyster breeding
and 198, 211–13, 216, 218, 238,
282, 286, 304, 325–6, 325n, 348,
361; plaster casts of fish 214–15,
215, 218–21, 220, 247, 252, 254,
263, 279, 282, 288, 364, 368–9,
369; pollution, identifies problem
of 128, 155, 213, 242, 243,
248–50, 252, 276–7, 278;